Biodiesel
Production and Properties

Biodiesel
Production and Properties

Amit Sarin
Amritsar College of Engineering and Technology, Amritsar, India
Email: amit.sarin@yahoo.com

RSC Publishing

ISBN: 978-1-84973-470-7

A catalogue record for this book is available from the British Library

Published by The Royal Society of Chemistry,
Thomas Graham House, Science Park, Milton Road,
Cambridge CB4 0WF, UK

Registered Charity Number 207890

For further information see our web site at www.rsc.org

Printed in the United Kingdom by Henry Ling Limited, Dorchester, DT1 1HD, UK

To
my brother
Dr Rakesh Sarin
I owe this work to you

Foreword

DR R. K. MALHOTRA
Director (R&D)

Indian Oil Corporation Ltd

The rapid growth and increasing aspirations of Society, particularly in developing economies like India and China continue to drive demand for petroleum fuels. Limited resources of fossil fuels have led to higher prices of oil and natural gas in recent years. In a country like India, where we depend largely on oil imports, it is imperative that we accelerate our search for alternative sources of energy. Biofuels, particularly from non-edible sources, hold promise but require major R&D efforts. Biodiesel is one such attractive option, which can not only be easily blended with diesel but would also provide significant environmental benefits.

The present book discusses all the major aspects of the production and properties of biodiesel. The book also covers the two major quality aspects of biodiesel; oxidation stability and low- temperature flow properties. The dependence of all of the major properties of biodiesel on its fatty acid composition is also discussed in detail. The chapter related to potential sources of biodiesel will be useful for researchers. To bring the book to a conclusion, the 'food *versus* fuel' issue is discussed, along with possible solutions. Overall the book covers every aspect of biodiesel from resource generation to utilization.

The book will be essential reading for students, chemists, chemical engineers and agricultural scientists working in both industry and academia. I congratulate and wish good luck to Dr Amit Sarin for this effort.

25 May 2012

Dr R. K. Malhotra

Biodiesel: Production and Properties
Amit Sarin
© Amit Sarin 2012
Published by the Royal Society of Chemistry, www.rsc.org

Preface

Why a new book?

In the course of my research and teaching, I refereed several research publications and books related to the field of biodiesels. I found that there is lot of information available on biodiesels, but the information is either scattered, or it does not cover every aspect of biodiesels. Therefore I felt the need for a reference as well as a textbook on biodiesels, which caters to the needs of students, teachers and researchers and covers the past, present and future aspects of biodiesels. In view of this, I was prompted to write a book entitled: *Biodiesel: Production and Properties.*

Contents of the book

This book has 10 chapters. Chapter 1 (*Introduction*) includes a brief history of biodiesel, such as when and who coined the term 'biodiesel'. In Chapter 2 (*Vegetable Oil as a Fuel: Can it be used Directly?*) different methods are discussed for reducing the viscosity of vegetable oils such as dilution (blending), micro-emulsification, pyrolysis (thermal cracking), and transesterification. The transesterification technique is used to minimize the viscosity and synthesize biodiesels. The technique is facilitated using catalysts including base homogenous and heterogeneous catalysts, acid homogenous and heterogeneous catalysts, acid–base heterogeneous catalysts and enzymes. The production of biodiesel using these catalysts with their advantages and disadvantages are analyzed elaborately. Other methods of biodiesel production like the BIOX co-solvent, non-catalytic supercritical alcohol, and catalytic supercritical methanol methods, as well as ultrasound and radio-frequency-

Biodiesel: Production and Properties
Amit Sarin
© Amit Sarin 2012
Published by the Royal Society of Chemistry, www.rsc.org

assisted and *in situ* biodiesel synthesis are discussed critically in detail. How biodiesel is purified and the influence of the different parameters on production, the molar ratio of alcohol to oil, the reaction temperature, water and free fatty acid contents, catalyst concentration and reaction time are discussed. The major advantages of biodiesel are also reported.

Chapter 3 (*Biodiesel Properties and Specifications*) includes many biodiesel properties, with their limits and methods. In Chapter 4 (*Oxidation Stability of Biodiesel*) the chemistry of the oxidation process is discussed. The factors affecting biodiesel stability are discussed in detail. The statistical relationship between oxidation stability and fatty acid methyl ester content is also discussed. The various methods of improving biodiesel stability, like the use of antioxidants, metal deactivators and the blending of biodiesels are discussed in detail. Chapter 5 (*Low- Temperature Flow Properties of Biodiesel*) deals in detail with the factors affecting the low- temperature flow properties of biodiesel and how these properties can be improved. The statistical relationships between low-temperature flow properties and fatty acid methyl ester content are included in this chapter. Chapter 6 (*Dependence of Other Properties of Biodiesel on Fatty Acid Methyl Ester Composition and Other Factors*) highlights the various prediction models for the properties of biodiesel like the viscosity, sulfur content, flash point, cetane number, carbon residue content, distillation temperature, water, free glycerol, contamination, glycerides, higher heating value, iodine value and density.

In Chapter 7 (*Diesel Engine Efficiency and Emissions using Biodiesel and its Blends*) the influence of biodiesel with other factors such as additives, on diesel engine performance and combustion and emission characteristics are discussed. Durability tests of diesel engines operated using biodiesel and its blends are also discussed. Statistical relationships between biodiesel performance and emission characteristics and fatty acid methyl ester composition are also discussed.

In Chapter 8 (*Major Resources for Biodiesel Production*) the possibilities of various feedstocks for biodiesel production with their detailed analyses are discussed. Other potential resources for biodiesel production are also discussed.

Chapter 9 (*Present State and Policies of the Biodiesel Industry*) highlights the present state and policies of the biodiesel industries in the EU, the USA, India and various other countries.

The production of biofuels from edible oils has raised concerns over preserving food security. In the last chapter (*The Food* Versus *Fuel Issue: Possible Solutions*), the possible solutions regarding this issue are discussed.

Acknowledgements

I would like to express my gratitude towards Mrs Janet Freshwater; Senior Commissioning Editor, Mrs Gwen Jones; former Commissioning Editor, Mrs Rosalind Searle; Commissioning Administrator, Ms Saphsa Codling; Books

Administrator, Ms Juliet Binns; Books Development Administrator, Dr Amaya Camara; Senior Production Controller (Books) and Ms Katrina Harding; Books Production Controller, at the Royal Society of Chemistry for their guidance during this project.

I would like to thank my parents Mr Madan Lal Sarin and Mrs Sudarshan Sarin for their blessings. I also thank my brothers, sisters-in-law, nephews and son for their support especially my wife Shruti for her help during the preparation of manuscript. I am short of words to thank my mentors Dr N. P. Singh, Dr R. K. Malhotra and Dr Rajneesh Arora. Thanks a lot again. I also thank all of my friends.

Above all, I thank God for blessing me to do this work.

I have tried my best to keep errors out of this book. I shall be grateful to the readers who point out any errors and make constructive suggestions.

Amit Sarin

Contents

Biodiesel: Production and Properties
Amit Sarin
© Amit Sarin 2012
Published by the Royal Society of Chemistry, www.rsc.org

List of Abbreviations

α –T	α- tocopherol
σ_{est}	Standard error of estimate
AAD	Average absolute deviation
AC	Air-cooled
ANN	Artificial Neural Network
AOCS	American Oil Chemist's Society
APE	Allylic position equivalent
ASA	American Soybean Association
ASABE	American Society of Agricultural and Biological Engineers
B100	Pure biodiesel
BAPE	Bisallylic position equivalent
BHA	Butylated hydroxyanisole
BHT	*tert*-butylated hydroxytoluene
Bmep	Brake mean effective pressure
BSEC	Brake-specific energy consumption
BSFC	Brake specific fuel consumption
BSU	Bosch Smoke Unit
BTDC	Before top dead center
BTE	Brake thermal efficiency
BXX	Biodiesel blends
CFPP	Cold filter plugging point
CHR	Cumulative heat release
CI	Compression Ignition
CN	Cetane number
COE	Carbon–oxygen equivalents
CP	Cloud point
CR	Carbon residue
CSFT	Cold soak filterability test

Biodiesel: Production and Properties
Amit Sarin
© The Royal Society of Chemistry 2012
Published by the Royal Society of Chemistry, www.rsc.org

CSMCRI	Central Salt & Marine Chemicals Research Institute
CSO	Cotton seed oil
CSOME	Cotton seed oil methyl ester
CVC	Calculated viscosity contribution
D	Density
DET	Differential export tax
DG	Diglycerides
DS	Direct standardization
EBB	European Biodiesel Board
EERL	Energy and Environment Research Laboratories
EGR	Exhaust gas regulation
EGT	Exhaust gas temperature
EIA	Energy Information Administration
EMA	Engine Manufacturer's Association
EMRA	Energy Market Regulatory Authority
FAEE	Fatty acid ethyl ester
FAME	Fatty acid methyl ester
FASOMGHG	Forest and the Agricultural Sector Optimization Model Greenhouse Gas
FBC	Fuel-borne catalyst
FFA	Free fatty acid
FP	Flash point
FT-IR	Fourier transform infrared
GC	Gas chromatography
GCVOL	Group-contribution volume
GHG	Greenhouse gas
GSP	Generalized System of Preferences
GWP	Global Warming Potential
HC	Hydrocarbons
HHV	Higher heating value
HOME	Honge oil methyl ester
HSDI	High-speed direct-injection
ID	Ignition delay
IDI	Indirect-injection
IFPRI	International Food Policy Research Institute
ILUC	Indirect land use change
IP	Induction period
IPCC	Intergovernmental panel on climate change
IPDT	Integral procedure degradation temperatures
IQT	Ignition quality tester
ITRI	Industrial Technology Research Institute
IV	Iodine value
JBD	Jatropha biodiesel
JME	Jatropha methyl ester
KME	Karanja methyl ester

LAME	Linoleic acid methyl ester
LFOR	Lean flame out region
LFR	Lean flame region
LHR	Low heat rejection
LTFT	Low-temperature flow test
LTVR	Low-temperature viscosity ratio
MD	Metal deactivator
ME	Mechanical efficiency
MEP	Mean effective pressure
MG	Monoglycerides
MHR	Maximum heat release rate
MHT	Maximum heat transfer
MME	Mahua methyl ester
MMT	Million metric tons
NA	Naturally aspirated
NBB	National Biodiesel Board
NBM	National Biodiesel Mission
NBRI	National Botanical Research Institute
ND	Normal diesel
NIR	Near infrared
NMR	Nuclear magnetic resonance
OAME	Oleic acid methyl ester
OBPA	Octylated butylated diphenyl amine
OS	Oxidation stability
OSI	Oil stability index
OX	Oxidizability
PAME	Palmitic acid methyl ester
PBD	Palm biodiesel
PCP	Peak cylinder pressure
PF	Poultry fat
PG	Propylgallate
PLS	Partial least squares
PM	Particulate matter
PME	Palm oil methyl ester
PoBD	Pongamia biodiesel
PP	Pour point
PuME	Pungam methyl ester
PY	Pyrogallol
R	Coefficient of regression
R^2	Coefficient of determination
RED	Renewable Energy Directive
RFS	Renewable fuel standard
RIN	Renewable identification number
RME	Rapeseed methyl ester
RSO	Rapeseed oil

RVOME	Recycled vegetable oil methyl ester
SE	Shell extra
SFO	Sunflower oil
SME	Soybean methyl ester
TBHQ	*tert*-butylhydroxquinone
TBP	*tert*-butylated phenol derivative
TCD	Total combustion duration
TG	Triglycerides
TME	Tallow methyl ester
TSOME	Tobacco seed oil methyl ester
ULSD	Ultra-low-sulfur diesel
V	Viscosity
WCO	Waste cooking oil
WCOME	Waste cooking oil methyl ester
WWTP	Wastewater treatment plant
YG	Yellow grease

CHAPTER 1

Introduction

1.1 Energy Demand

Today, more and more countries are prospering through economic reforms and becoming industrially advanced. Energy is a basic requirement for the economic development of every country. Every sector of the economy, including the agriculture, industry, transport, commercial and domestic sectors require energy. According to a report available on the United States Energy Information Administration (EIA) website, total world energy consumption was 406 quadrillion British thermal units (Btu) in 2000 and is projected to increase to 769.8 quadrillion Btu by 2035 (Figure 1.1).[1] This is an increase of approximately 47.25%, and energy consumption will definitely increase further.

Can you imagine a world without the current mainstream fuels such as diesel or petrol used to run your vehicles, to cook your the food, to run every sector of the economy and so many other things where these fuels are required? The obvious answer is no. But the next question that arises here is for how long will this fuel be there to serve the needs of mankind? Are these fuels never going to end? No, the supply of these fuels is decreasing for so many obvious reasons. But what will happen if these fuels are not sufficient to cater for the needs of mankind? What will happen if crude oil prices will reach beyond the range of most people? We all are intentionally, consciously or subconsciously aware that day will come.

Therefore there is a need to explore alternative energy resources and the research community is taking part in this. Mainstream forms of renewable energy include wind power, hydropower, solar energy, biomass, and biofuels. The contribution of all these resources is important because of the aforementioned reasons, and biodiesel could be one of the major contributors.

Biodiesel: Production and Properties
Amit Sarin
© Amit Sarin 2012
Published by the Royal Society of Chemistry, www.rsc.org

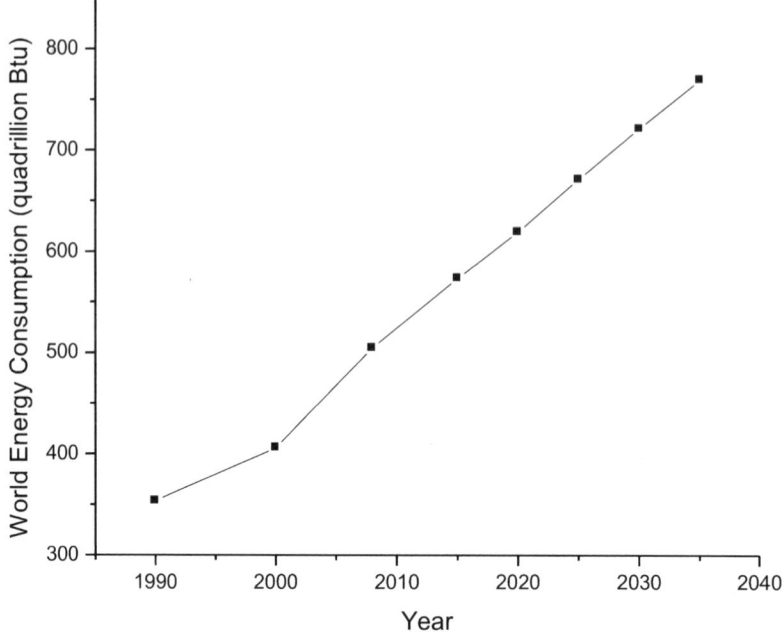

Figure 1.1   Total world energy consumption (1990–2035).

1.2   Biodiesel: Green Fuel, Fuel of the Future or Magic Fuel?

Diesel fuels play an important role in the industrial development of most countries. These fuels play a major part in the transport sector and their consumption is steadily increasing. The intensity of fuel consumption is directly proportional to the development of the society. Diesel engines have been widely used in engineering machinery, automobiles and shipping equipment because of their excellent drivability and thermal efficiency. Diesel fuels are used in heavy trucks, city transport buses, locomotives, electricity generators, farm equipment, underground mining equipment *etc*.

The energy generated from the combustion of fossil fuels has indeed enabled many technological advancements and social–economic growth. However, it has simultaneously created many environmental concerns, which can threaten the sustainability of our ecosystem. The high demand for diesel in the industrialized world and pollution problems caused by its widespread use make it necessary to develop renewable energy sources of limitless duration with a smaller environmental impact than these traditional sources.

Therefore, to replace diesel fuel another renewable fuel is required and that could be biodiesel. Biodiesel is defined technically as 'a fuel comprised of monoalkyl esters of long-chain fatty acids derived from vegetable oils or animal fats, designated B100, and meeting the requirements of the American Society for Testing and Materials D-6751'.[2]

The importance of the use of vegetable oil as a diesel engine fuel was cited by none other than Sir Rudolf Diesel. Speaking to the Engineering Society of St. Louis, Missouri, in 1912 he said, "The use of vegetable oils for engine fuels may seem insignificant today, but such oils may become in the course of time as important as petroleum and the coal tar products of the present times".[3]

The first known report of using esters of vegetable oils as a motor fuel was described in a Belgian patent granted on 31 August 1937.[4] Following this, there was a report on using esters in Brazil.[5] A conference entitled 'Vegetable Oil Fuels' was held in Fargo, North Dakota, in August 1982, under the auspices of the American Society of Agricultural Engineers (ASAE; now the American Society of Agricultural and Biological Engineers, ASABE).[6] The conclusion of the conference was that raw vegetable oils can be used as fuels but can lead to problems such as injector coking, polymerization in the piston ring belt area causing stuck or broken piston rings, and a tendency to thicken lubricating oil causing sudden and catastrophic failure of the rod and/or crankshaft bearings. A method for reducing the viscosity of the oil must be developed and the most likely technique for that was transesterification of vegetable oil (discussed in detail in Chapter 2). After that, various researchers worked to develop and search various resources for the production of biodiesel (discussed later in this book).

1.2.1 When and Who Coined the Term 'Biodiesel'?

While searching for the answer to the question when and who coined the term biodiesel, I found an article entitled 'Biodiesel: An alternative fuel for compression ignition engines'.[7] In the article, the authors mention that they found a flyer from Bio-Energy (Australia) Pty Ltd that promotes equipment to produce a 'Low Cost Diesel Fuel' called 'Bio-Diesel', but the flyer was not dated. The flyer was found attached to a letter that uses the word 'Bio-Diesel' and the letter was also not dated. The authors had found three other articles dated 1984, one of which stated "...the two fuels, BioDiesel and distillate, are virtually identical...".[8] Therefore it can be concluded that the term 'biodiesel' was coined around 1984, however who actually used the term first is still doubtful.

Can there be any fuel which can act like magic? Can biodiesel be that fuel? Can biodiesel be the fuel of the future? In this book, I have tried to explore everything related to biodiesel: how biodiesel came into the picture; how it is synthesized; what its main properties are; the various sources of biodiesel the current status of the biodiesel industry and solutions to the food *versus* fuel issue. For me, biodiesel could be a magic fuel or the fuel of the future if explored properly. We shall see what your thoughts are after reading this book.

References

1. http://www.eia.gov/forecasts/ieo/pdf/0484(2011).pdf.
2. http://www.biodiesel.org/what-is-biodiesel/biodiesel-basics.
3. http://www.biodiesel.org/news/biodiesel-news/news-display/2012/03/18/recognizing-rudolf-diesel's-foresight-in-celebrating-biodiesel-day.
4. G. Chavanne, *Belg. Pat.*, 422877, 1937.
5. G. H. Pischinger, A. M. Falcon, R. W. Siekman, and F. R. Fernandes, presented at the 5th International Symposium on Alcohol Fuel Technology, Auckland, 1982.
6. American Society of Agricultural Engineers, presented at the International Conference on Plant and Vegetable Oil as Fuels, St. Joseph, 1982.
7. J. H. Van Gerpen, C. L. Peterson and C. E. Goering, presented at the Agricultural Equipment Technology Conference, Louisville, 2007.
8. *Alternate fuel could save money*, Power Farming Magazine, 10 Oct. 1984, 93(10).

CHAPTER 2

Vegetable Oil as a Fuel: Can it be used Directly?

2.1 Introduction

The possibility of using vegetable oil as a fuel has been recognized since the beginning of diesel engines. Researchers have searched for alternate fuel sources and concluded that vegetable-oil-based fuels can be used as alternatives.[1] Speaking to the Engineering Society of St Louis, Missouri in 1912, Rudolph Diesel, said, "The use of vegetable oils for engine fuels may seem insignificant today, but such oils may become in the course of time as important as petroleum and the coal tar products of the present times".[2] During the 1930s and 1940s, particularly during World War II, vegetable oils were used in emergencies as substitutes for diesel.[3,4]

Vegetable oils have higher viscosities than commercial diesel fuel.[5] The high viscosity of raw vegetable oil reduces fuel atomization and increases penetration, which is partially responsible for engine deposits, piston ring sticking, injector coking and the thickening of oil.[6,7] Different methods have been developed to reduce the viscosity of vegetable oils such as dilution (blending), micro-emulsification, pyrolysis (thermal cracking), and transesterification.

2.2 Dilution (Blending)

Crude vegetable oils can be blended directly or diluted with diesel fuel to improve their viscosity. Dilution reduces the viscosity and engine performance problems such as injector coking and the creation of carbon deposits. In 1980,

Biodiesel: Production and Properties
Amit Sarin
© Amit Sarin 2012
Published by the Royal Society of Chemistry, www.rsc.org

Caterpillar Brazil used a 10% mixture of vegetable oil and diesel to maintain total power without any modification or adjustment to the engine. A diesel engine study with a blend of 20% vegetable oil and 80% diesel fuel has also been carried out.[8] Twenty-five percent sunflower oil and 75% diesel were blended as a diesel fuel and the reduced viscosity was 4.88 cSt at 313 K, while the maximum specified American standard test method (ASTM) value is 4.0 cSt at 313 K.[9] This mixture was not suitable for long-term use in a direct-injection engine. The viscosity decreases with increasing percentage of diesel. Further, it was also reported that the viscosity of a 25 : 75 high oleic sunflower oil : diesel fuel blend was 4.92 cSt at 40 °C and that it has passed the 200 h Engine Manufacturers Association (EMA) test. Another study was conducted by using the blending technique on frying oil.[10]

2.3   Micro-emulsification

Micro-emulsification is another approach to reducing the viscosity of vegetable oils. A micro-emulsion is defined as a colloidal equilibrium dispersion of an optically isotropic fluid microstructure with dimensions generally in the 1–150 nm range formed spontaneously from two normally immiscible liquids and one or more ionic amphiphiles.[11] In other words; micro-emulsions are clear, stable isotropic fluids with three components: an oil phase, an aqueous phase and a surfactant. The aqueous phase may contain salts or other ingredients, and the oil may consist of a complex mixture of different hydrocarbons and olefins. This ternary phase can improve spray character-istics by explosive vaporization of the low-boiling-point constituents in the micelles. All micro-emulsions with butanol, hexanol and octanol meet the maximum viscosity limitation for diesel engines.[12] A micro-emulsion prepared by blending soybean oil, methanol, 2-octanol and a cetane improver in the ratio of 52.7 : 13.3 : 33.3 : 1.0 has passed the 200 h EMA test.[13]

2.4   Pyrolysis (Thermal Cracking)

Pyrolysis is a method of conversion of one substance into another through heating or heating with the aid of a catalyst in the absence of air or oxygen.[14] It involves cleavage of chemical bonds to yield small molecules.[15] The material used for pyrolysis can be vegetable oils, animal fats, natural fatty acids and methyl esters of fatty acids. The liquid fuel produced by this process has an almost identical chemical composition to conventional diesel fuel.[16] Soybean oil has been thermally decomposed in air using the standard ASTM method for distillation. The viscosity of the pyrolyzed soybean oil distillate is 10.2 cSt at 37.8 °C, which is higher than the ASTM specified range for No. 2 diesel fuel but acceptable as it is still well below the viscosity of soybean oil.[17]

2.5 Biodiesel and its Production: Transesterification

Transesterification, also called alcoholysis, is a chemical reaction of an oil or fat with an alcohol in the presence of a catalyst to form esters and glycerol. It involves a sequence of three consecutive reversible reactions where triglycerides (TGs) are converted to diglycerides (DGs) and then DGs are converted to monoglycerides (MGs) followed by the conversion of MGs to glycerol. In each step an ester is produced and thus three ester molecules are produced from one molecule of TG.[18] Among the alcohols that can be used in the transesterification reaction are methanol, ethanol, propanol, butanol and amyl alcohol. Methanol and ethanol are used most frequently. However methanol is preferred because of its low cost. Figure 2.1 shows the transesterification reaction of TGs with alcohol. A catalyst is usually used to improve the reaction rate and yield. Because the reaction is reversible, excess alcohol is used to shift the equilibrium to the product side. It also produces glycerol as a byproduct which has some commercial value. Here R_1, R_2, R_3 are long-chain hydrocarbons (HC), sometimes called fatty acid chains. Normally, there are five main types of chains in vegetable oils and animal oils: palmitic, stearic, oleic, linoleic, and linolenic. When the TG is converted stepwise to a diglyceride, then a monoglyceride, and finally to glycerol, one mole of fatty ester is liberated at each step (Figure 2.2).[19] All of the steps in biodiesel production are shown in Figure 2.3.

Transesterification is the most viable process adopted known so far for the lowering of viscosity and for the production of biodiesel. Thus biodiesel is the alkyl ester of fatty acids, made by the transesterification of oils or fats, from plants or animals, using short-chain alcohols such as methanol and ethanol in

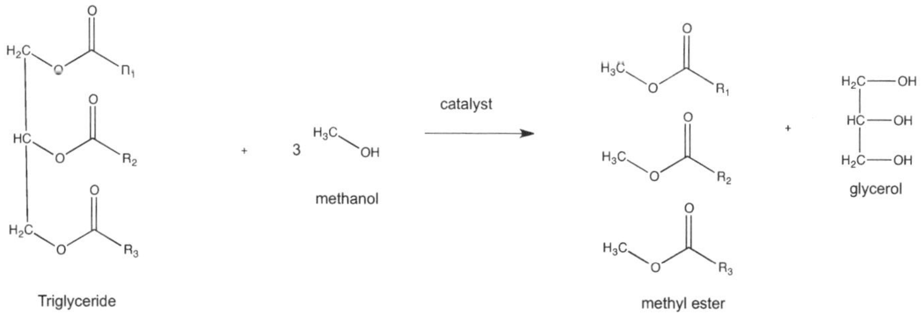

Figure 2.1 Transesterification of triglycerides with alcohol.

Triglyceride   + ROH ↔ Diglyceride   + $RCOOR_1$

Diglyceride   + ROH ↔ Monoglyceride + $RCOOR_2$

Monoglyceride   + ROH ↔ Glycerol   + $RCOOR_3$

Figure 2.2 The steps in transesterification.

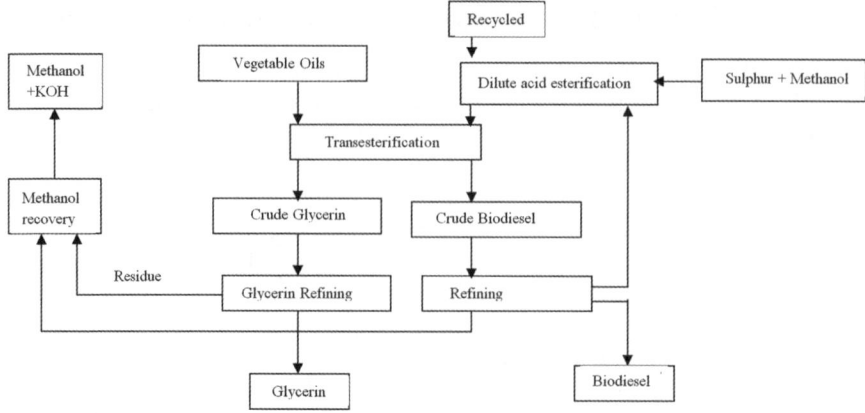

Figure 2.3   Scheme for biodiesel production.

the presence of a catalyst. Glycerin is consequently a by-product of biodiesel production (Figure 2.1). Pure biodiesel or 100% biodiesel is referred to as 'B100'. A biodiesel blend is pure biodiesel blended with petrodiesel. Biodiesel blends are referred to as BXX. The XX indicates the amount of biodiesel in the blend (*i.e.*, a B90 blend is 90% biodiesel and 10% petrodiesel).

2.6   Use of Catalysts in Transesterification

The catalysts used in the transesterification are broadly divided into two types: (a) Base catalysts and (b) acid catalysts.

Base catalysts are preferred over acid catalysts, due to their capability of completion of reaction at higher speed, their requirement for lower reaction temperatures, and their higher conversion efficiency as compared to acid catalysts. Researchers have suggested that base catalysis is successful only when the free fatty acid (FFA) is less than one percentage, but it was also observed that base catalysts can be used in cases where the FFA content is greater than one, but more catalyst is needed.[19,20] Later on, it was advocated that base catalysts exhibit excellent results when the FFA content of the oil is below two. It has also been reported that the rate of the transesterification reaction becomes a thousand times faster when a base catalyst is used instead of an acid catalyst.[21] But these are not successful for oils having FFA content greater than three.[22] However, it has also been reported that base catalysts could be effectively used for feedstock having FFA contents up to five.[23]

The base catalysts cause saponification when they react with FFAs present in the vegetable oils or triglyceride, particularly when the acid value of the feedstock is high.[24] In such cases, acid catalysts are used. The acid value of edible oils is normally low compared to non-edible oils. However, the acid value of edible oils also increases when they are used for frying purpose for long periods.[25-27] In such cases the use of acid catalysts shows better results.

2.6.1 Base Catalysts

The reaction mechanism for base-catalyzed transesterification is shown in Figure 2.4.[28] The reaction mechanism has three steps. The first step is the reaction of the carbonyl carbon atom with the anion of the alcohol, forming a tetrahedral intermediate, from which the alkyl ester and corresponding anion of the DG are formed. Another catalytic cycle is started when the catalyst reacts with a second molecule of alcohol. From there, DGs and MGs are converted into alkyl esters and glycerol.[28,29]

2.6.1.1 Homogeneous Base Catalysts

If the catalyst used in the reaction remains in the same (liquid) phase to that of the reactants during transesterification, it is homogeneous catalytic transesterification. Various types of homogeneous base catalysts are used for transesterification. The most common among these are sodium hydroxide, sodium methoxide, potassium hydroxide, and potassium methoxide.

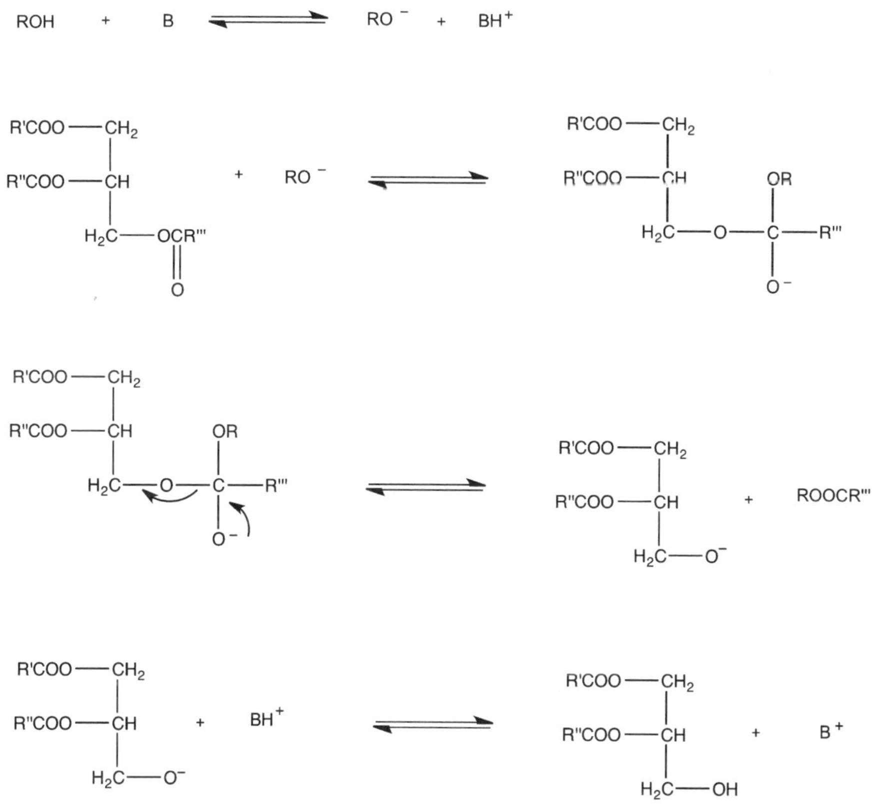

Figure 2.4 Reaction mechanism of base-catalyzed transesterification (B: base).

Sodium hydroxide (NaOH). The use of NaOH as a catalyst is preferred over potassium hydroxide because it causes less emulsification, eases the separation of glycerol, and is of lower cost. There are various reports of use of NaOH as a catalyst for biodiesel production.[30–39]

Potassium hydroxide (KOH). KOH is a base catalyst which is widely used in the transesterification process. The performance of KOH was better than that of NaOH and the separation of biodiesel and glycerol was easier when KOH was used as a catalyst; hence it was preferred over NaOH.[40–43]

Sodium methoxide (NaOCH$_3$). NaOCH$_3$ is more effective than sodium hydroxide as a catalyst because it disintegrates into CH_3O^- and Na^+ and does not form water in contrast to NaOH/KOH. Moreover only 50% of it is required compared to NaOH. But the catalyst is less common due to its higher cost. It was found that 0.5% sodium methoxide and 1% sodium hydroxide exhibited similar results with methanol-to-oil molar ratios of 6.[44] The use of NaOCH$_3$ is reported in more of the literature.[45,46]

Potassium methoxide (CH$_3$OK). CH$_3$OK is a base catalyst which can also be used for transesterification. Although it has been tested by many researchers, very few have recommended using it on regular basis.[47–49]

2.6.1.2   Heterogeneous Base Catalysts

If the catalyst remains in a different phase (*i.e.*, solid, immiscible liquid or gaseous) to that of the reactants the process is called heterogeneous catalytic transesterification. The heterogeneous catalytic transesterification is included under the term 'Green Technology' for the following reasons: (1) the catalyst can be recycled, (2) there is no or very little waste water produced during the process, and (3) separation of the biodiesel from the glycerol is much easier.

Homogeneous catalysts are very effective catalysts but the major problem associated with the use of these catalysts is their removal from the methyl ester requiring excessive washing. Therefore, a lot of energy, water, and time are consumed; moreover these catalysts cannot be reused. In contrast, hetero-geneous (solid) base catalysts being insoluble, are separated simply by filtration and can be reused many times. A lot of research has also been done in this direction. Commonly used solid base catalysts are alkaline earth metal oxides, zeolites, KNO$_3$ loaded on Al$_2$O$_3$, KNO$_3$/Al$_2$O$_3$, BaO, SrO, CaO and MgO.[50,51]

Alkaline earth metal oxides and derivatives. Alkaline earth metals such as Mg, Ca, Sr, Ba and Ra, their oxides and derivatives are used by different researchers. MgO is widely used among the other alkaline earth metals, which all have good heterogeneous natures as catalysts.[52–61] Among the alkaline earth metal oxides CaO is the most widely used as a catalyst for transesterification and a 98% fatty acid methyl ester (FAME) yield during the first cycle of reaction has been reported.[62] The reactivity of CaO is further determined by its calcination temperature. However, the reusability of the catalyst for further steps remains in question. Modification of CaO to make it organometallic in nature, *e.g.*, $Ca(OCH_3)$ or $Ca(C_3H_7O_3)_2$, however, has been very effective with respect to reusability. $Ca(C_3H_7O_3)_2/CaCO_3$ has also been observed to be an efficient heterogeneous catalyst with reusability for five cycles and FAME yields as high as 95%.[63] CaO slightly dissolves in methanol and the soluble substance can be identified as calcium diglyceroxide, in which CaO reacts with glycerol during the transesterification of soybean oil with methanol.[64] SrO can also be used as a heterogeneous catalyst for transesterification.[52,61,63] However very limited literature is available, as discussed below. The reusability of the catalyst was reported to be 10 times.

Boron and carbon group elements. Boron group elements particularly aluminum and Al_2O_3 loaded with various other metal oxides, halides, nitrates and alloys are used for biodiesel production.[65–68] Carbon-based catalysts are easy to prepare and economically feasible heterogeneous catalysts.[69,70]

Waste-material-based heterogeneous catalysts. There are many natural calcium sources such as eggshell, mollusk shell, bones *etc.* which are widely used as raw materials for catalyst synthesis and could eliminate waste whilst simultaneously producing catalysts with high cost effectiveness. CaO catalysts derived from these waste materials could be a potential candidate for biodiesel production.[71–73]

Biodiesel production using different metal oxides and derivatives. The most common transition metals and their oxides used for biodiesel production are *viz.* ZnO, TiO, TiO_2/SO_4^{2-} and ZrO_2/SO_4^{2-}, ZnO and ZrO as base heterogeneous catalysts.[56,61,74–76] The activity of zirconium oxide functionalized with tungsten oxide and sodium molybdate (Na_2MoO_4) was also studied.[77,78] There are also reports of usage of mixed metal oxides with their derivatives.[79,80]

Zeolite-based catalysts. Zeolites as catalysts have the characteristics of acidic sites and shape selectivity. Zeolites vary in pore structure, and inner electric fields from their crystal and surface properties contribute to their varying catalytic properties. Zeolites can accommodate a wide variety of cations such as Na^+, K^+, Ca^{2+}, Mg^{2+} and many others, which contribute to their basic nature. Zeolites as potential heterogeneous catalysts for the preparation of biodiesel have been investigated.[51,81–84]

2.6.2 Acid Catalysts

Although base catalysts are very efficient and popular for transesterification, these catalysts do not exhibit good results when the feedstock contains water and its acid value is high. Base catalysts are highly sensitive to water content, as this causes soap formation, making separation difficult. The acid values of most of the non-edible oils are higher than the performance range of base catalysts. Thus, in such cases, acid catalysts are used. The 'acid value' represents the number of acidic functional groups and is measured in terms of the quantity of potassium hydroxide required to neutralize the acidic characteristics of the sample. The mechanism of the acid-catalyzed transesterification of vegetable oil is shown in Figure 2.5. The protonation of the carbonyl group of the ester promotes the formation of a carbocation, which after nucleophilic attack of the alcohol produces a tetrahedral intermediate.[85] This intermediate will eliminate glycerol to form a new ester and to regenerate the catalyst. Acid-catalyzed transesterification should be carried out in the absence of water.[28]

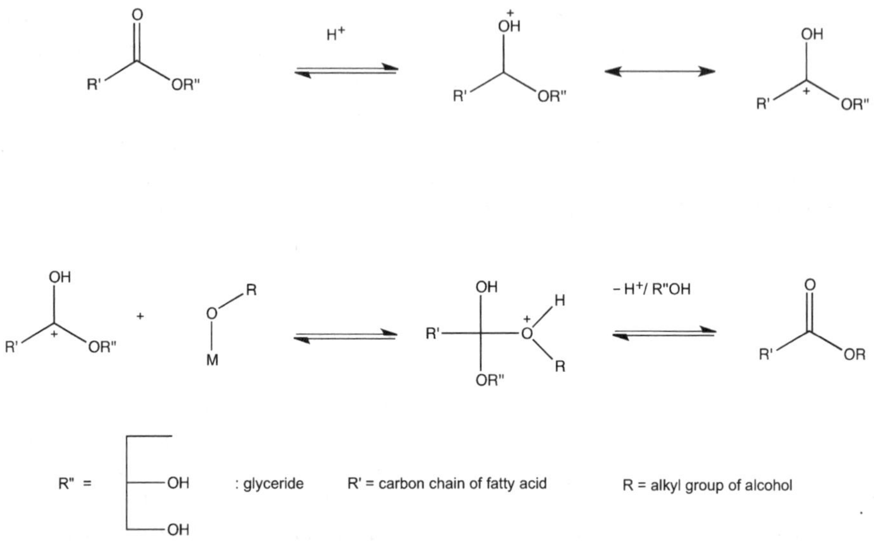

Figure 2.5   Reaction mechanism of acid-catalyzed transesterification.

The problems associated with these catalysts are: the requirement for more alcohol; higher reaction temperatures and pressures; slower reaction rates; reactor corrosion and environmental issues.[30,86]

Both homogeneous and heterogonous acid catalysts can be used for transesterification. The acid catalysts more commonly used include, sulfuric acid, hydrochloric acid, phosphoric acid, and sulfonated organic acids. The FFA content of neat edible oils is normally low but these oils are costly and conversion of too much edible oil into biodiesel may cause food crises. Therefore the only choice is the use of waste oils or non-edible feedstock. The FFA content of non-edible oils is generally high. The FFA content of edible oils is increased when these oils are used for frying purposes, due to hydrolysis of triglycerides. Acid catalysts are recommended for handling such feedstock.

Pre-treatment of acidic feedstock. Acid catalysts are generally used for two-step transesterification. In the first step the oils react with alcohol in the presence of the acid catalyst. The acid value of the product is reduced and the oil re-reacted with methanol in the presence of a base catalyst.[34,87–90] This pre-treatment has the effect of decreasing the FFA content to the alkaline transesterification range. Generally the value is reduced to less than one. In the next step the oil is re-transesterified using a catalyst.[91]

Many pre-treatment methods have been proposed for reducing the high FFA content of the oils, including steam distillation, extraction by alcohol, and esterification by acid catalysis.[92–94] However, steam distillation requires a high temperature and has low efficiency. Due to the limited solubility of FFAs in alcohol, extraction by the alcohol method requires a large amount of solvent and the process is complicated.

As discussed earlier, FFAs will be converted to biodiesel by direct acid esterification after which the water needs to be removed. If the acid value of the oils or fats is very high, a one-step esterification pre-treatment may not reduce the FFA content efficiently because of the high content of water produced during the reaction. In this case, a mixture of alcohol and sulfuric acid can be added to the oils or fats three times. The time required for this process is about 2 h and water must be removed by a separating funnel before adding the mixture into the oils or fats for re-esterification.[95] Moreover, some researchers reduce the percent of FFAs by using acidic ion-exchange resins in a packed bed. Strong commercial acidic ion-exchange resins can be used for the esterification of FFAs in waste cooking oils but the loss of catalytic activity maybe a problem.[96–100]

An alternative approach to reduce the FFA content is to use iodine as a catalyst to convert FFA into biodiesel. The advantage of this approach is that the catalyst can be recycled after the esterification reaction. It was found through orthogonal tests, under optimal conditions (*i.e.*, iodine amount: 1.3 wt% of the oil; reaction temperature: 80 °C; ratio of methanol to oil: 1.75 : 1; reaction time: 3 h) that the FFA content can be reduced to <2%.[101]

Another new method of pre-treatment is to add glycerol to the acidic feedstock and heat it to a temperature of about 200 °C, normally with a catalyst such as zinc chloride.[102] Then the FFA level will reduce and biodiesel can be produced using the traditional alkali-catalyzed transesterification method. The advantage of this approach is that no alcohol is needed during the pre-treatment and the water formed during the reaction can be immediately vaporized and vented from the mixture. However, the drawbacks of this method are its high temperature requirement and relatively slow reaction rate.

2.6.2.1 *Homogeneous Acid Catalysts*

Researchers have converted feedstocks with higher acid values into biodiesel using homogeneous acid catalysts through a two-step esterification.[20,103–108] In the first step they used liquid acid and methanol for the esterification, reduced the acid value and re-esterified the oils using a base catalyst. Although acid catalysts are cheaper than base catalysts, more alcohol is needed when acid catalysts are used. Moreover, in the presence of acids, reactors made of special materials are required; therefore the overall cost is increased.[109]

2.6.2.2   Heterogeneous Acid Catalysts

Heterogeneous acid catalysts are preferred over homogeneous catalysts, because they do not dissolve in the alcohol and feedstock, thus they can be separated easily by filtration and can be re-used. Such catalysts are effective for the esterification of FFA as well as of TGs.[110]

Currently, biodiesel research is focused on exploring new and sustainable solid acid catalysts for transesterification reaction. In addition, it is believed that solid acid catalysts have the strong potential to replace liquid acid catalysts.[111] The advantages of using solid acid catalysts are: (a) they are insensitive to FFA content; (b) esterification and transesterification occurs simultaneously; (c) they eliminate the washing step of the biodiesel; (d) easy separation of the catalyst from the reaction medium is possible, resulting in lower product contamination level; (e) easy regeneration and recycling of the catalysts and (f) a reduction in corrosion problems, even in the presence of the acid species.[112,113]

The development of heterogeneous catalyst systems is an important factor to be incorporated into continuous-flow reactors because a continuous process can minimize product separation and purification costs, making it economically viable and able to compete with commercial petroleum-based diesel fuels.[114,115] The ideal solid acid catalyst for transesterification should have characteristics such as an interconnected system of large pores, a moderate-to-high concentration of strong acid sites, and a hydrophobic surface.[112]

Research on the direct use of solid acid catalysts for biodiesel production has not been widely explored due to its limitations of slow reaction rate and possible undesirable side-reactions. The following section will give an overview of the various solid acid catalysts reported so far for biodiesel production.

Zirconium oxide (ZrO$_2$), sulfated oxides and cation-exchange resins. There have been several studies on ZrO$_2$ as a heterogeneous acid catalyst for the transesterification of different feedstocks due to its strong surface acidity. SO$_4^{2-}$/ZrO$_2$ could produce promising results in transesterification with yields reaching as high as 90.3%.[74] However, when unsulfated ZrO$_2$ was used as a catalyst instead of SO$_4^{2-}$/ZrO$_2$, only a 64.5% yield was reported. This indicates that modification of the metal oxide surface acidity is the key factor in obtaining high conversion of TGs. Apart from that, combination of alumina, Al$_2$O$_3$ with ZrO$_2$ and modification of ZrO$_2$–Al$_2$O$_3$ with tungsten oxide (WO$_3$) not only provides high mechanical strength but also enhances the acidity of the catalyst.[111]

The performance of tungstated zirconia–alumina (WZA) and sulfated zirconia–alumina (SZA) in transesterification with methanol at 200–300 °C using a fixed bed reactor under atmospheric pressure has been reported.[116] WZA was found to have a higher activity in transesterification as compared with SZA although the causes of the improved activity of the WZA catalyst was not reported.

A relatively low ester yield of 65% was obtained when the transesterification reaction was carried out at a lower reaction temperature of 200 °C and for a shorter reaction time of 10 h.[117] It was found that WO_3/ZrO_2 has a higher stability than SO_4^{2-}/ZrO_2, therefore avoiding the leaching of acid sites into the reaction media.[118] Even if WO_3 leached into the reaction media, it would not contaminate the product. In the study, it was found that 85% of FFA conversion was attained in a packed-bed reactor after 20 h of reaction time at 75 °C, but this decreased to 65% and remained stable thereafter. The reason given was due to the oxidation of WO_3 after long-term exposure to FFA (a reducing agent) that resulted in a decrease in catalytic activity. Therefore, leaching of WO_3 was ruled out as the main reason for catalyst deactivation. In addition, WO_3/ZrO_2 could be simply regenerated by air re-calcination. However, further study on the catalysis, process optimization and also the oxidation state of WO_3 is still required.[114,119]

Sulfated zirconia (SZ) was used in a solvent-free method to prepare biodiesel by a two-step process of esterification and transesterification.[120] An acid catalyst was used for the esterification and an alkali catalyst (KOH) was used for the transesterification reaction. FFA conversion was achieved using 1 wt% SZ as an acid catalyst with a methanol-to-oil molar ratio of 9 : 1, a temperature of 65 °C and reaction time of 2 h. The acid value was reduced to 94% of the raw oil (24.76 mg KOH per g), which confirmed the conversion. Therefore, this pre-treatment reduces the overall complexity of the process and a conversion efficiency of 95% was reported when pre-treated oil was reacted with methanol in the presence of KOH.

There is also number of reports of the usage of sulfated oxides as acid heterogeneous catalysts for biodiesel production.[117,121–124] A novel organic–inorganic hybrid membrane as a heterogeneous acid catalyst for biodiesel production from zirconium sulfate ($Zr(SO_4)_2$) and sulfonated poly(vinyl alcohol) (SPVA) was synthesized. Esterification results showed that the conversions of FFA in acidified oil were 94.5% and 81.2% for $Zr(SO_4)_2S/PVA$ and $Zr(SO_4)_2/PVA$ catalytic membranes, respectively. The stability of the $Zr(SO_4)_2/SPVA$ catalytic membrane is superior to the $Zr(SO_4)_2/PVA$ catalytic membrane. An efficient microwave-assisted transesterification (MAT) technique to prepare biodiesel from yellowhorn (*Xanthoceras sorbifolia Bunge.*) oil with a heteropolyacid (HPA) catalyst namely $Cs_{2.5}H_{0.5}PW_{12}O_{40}$ has also been reported.[124] The maximum yield of FAME reached 96.22% under optimal conditions: temperature 60 °C; reaction time 10 min; molar ratio of methanol to oil 12 : 1; 1% (w/w of oil) catalyst and minimum recyclability nine times. The results showed that the $Cs_{2.5}H_{0.5}PW_{12}O_{40}$ heterogeneous acid catalyst had a higher efficiency for transesterification under microwave irradiation as compared with the conventional method. Cation-exchange resins are also reported extensively for laboratory-scale biodiesel preparation although industrial application has not yet been explored.[125–132]

2.6.3   Acid–Base Heterogeneous Catalysts

Heterogeneous catalysts having both acidic and basic sites have also been investigated which could simultaneously esterify FFA and transesterify TGs to biodiesel. A microwave-assisted transesterification reaction was performed in the presence of methanol or ethanol, using an alcohol-to-castor bean oil molar ratio of 6 : 1, and 10% (w/w) acidic silica gel or basic alumina (in relation to the oil mass) as the catalyst.[133] Using acid catalysis, the reaction occurred with satisfactory yields using H_2SO_4 immobilized in SiO_2, and methanol at a temperature of 60 °C for 3 h as well as using microwave irradiation for 30 min. The best results were obtained under basic conditions (Al_2O_3/50% KOH) using methanol and conventional (60 °C, stirring, 1 h) or microwave conditions (5 min). In comparison with conventional heating, the catalyzed alcoholysis assisted by microwaves is much faster and leads to higher yields of the desired fatty esters. Various researchers have investigated these types of catalyst for biodiesel production.[74,134–138]

2.6.4   Enzyme Catalysts

The transesterification of vegetable oils or animal fats is also possible by using enzyme catalysts or biocatalysts. This is preferred because, during their use, soap is not formed; therefore there is no problem of purification, washing, and neutralization. Another advantage of using enzymatic catalysts is that the reactions can be carried under at mild conditions. Enzymatic catalysts can also be applied to feedstocks of high FFA and convert more than 90% of oil into biodiesel. However, there are also problems associated with enzyme catalysts like their higher cost and longer reaction times.[139] Enzyme catalysts such as lipozyme, novozym 435, lipozyme TLIM and lipozyme RMIM have shown good results for biodiesel production.[140–142] Three different enzymes (i) *Chromobacterium viscosum*, (ii) *Candida rugosa*, and (iii) Porcine pancreas were used for transesterification and *Chromobacterium viscosum* was the best catalyst, returning a 71% yield in 8 h.[143] Lipase from *Pseudomonas fluorescens* could yield more than 90% biodiesel.[144] There are also other reports of the use of enzyme catalysts for biodiesel production.[145–149]

2.7   Other Methods of Biodiesel Production

2.7.1   BIOX Co-Solvent Method

BIOX is a technology development company which is a joint venture between the University of Toronto Innovations Foundation and Madison Ventures Ltd. The BIOX co-solvent method was developed by Boocock *et al.* in 1996.[150] In this process, TGs are converted to esters through the selection of inert co-solvents that generate a one-phase oil-rich system.[102,150,151] Co-solvent is used to overcome the slow reaction times caused by the extremely low solubility of the alcohol in the triglyceride phase. Tetrahydrofuran (THF) as a co-solvent was used to make the methanol soluble.[151] After completion of the reaction, the biodiesel–glycerol phase separation was cleaned and both the excess alcohol and the THF co-solvent were recovered in a single step. However, due to the possible hazard and toxicity of the co-solvents, they must be completely removed from the glycerol phase as well as the biodiesel phase and the final products should be water-free.[150] The main advantage of the BIOX co-solvent process is that it uses inert, reclaimable co-solvents in a single-pass reaction that takes only seconds at ambient temperature and pressure, and no catalyst residues appear in either the biodiesel phase or the glycerol phase.[102] This process can handle not only grain-based feedstocks but also waste cooking oils and animal fats. It was however found that the recovery of excess alcohol is difficult when using this process because the boiling point of the THF co-solvent is very close to that of methanol.[102,152]

2.7.2 Non-Catalytic Supercritical Alcohol Transesterification

Supercritical methanol transesterification is a method through which the vegetable oils or animal fats can be converted into biodiesel in a highly reduced reaction time without the use of any type of catalyst. The feedstock is reacted with supercritical methanol at extremely high pressure and temperature. Due to the absence of catalyst, no washing and neutralization is required. Another advantage of this process is that the water content does not affect the reaction. However, there are also problems associated with this method such as the requirement of high pressure and temperature resulting in the high cost of the apparatus. The reported time, temperature and pressure vary from 222 s to 6 min, 222 to 400 °C, and 80 to 650 bars respectively for 95 to 100% yield of biodiesel.[150,153–156]

2.7.3 Catalytic Supercritical Methanol Transesterification

Catalytic supercritical methanol transesterification is performed in an autoclave in the presence of 1–5% NaOH, CaO, and MgO as a catalyst at 520 K. The yield of conversion rises to 60–90% for the first minute.[157] Transesterification of rapeseed oil with supercritical/subcritical methanol in the presence of a relatively low amount of NaOH (1%) was successfully carried out, without the occurrence of soap formation.[158]

2.7.4 Ultrasound- and Radio-Frequency-Assisted Transesterification

Ultrasound is defined as sound of frequency within the range 20 kHz to beyond 100 MHz and it has proven to be a useful tool in enhancing reaction rates in many reaction systems. During transesterification, ultrasonication provides the mechanical energy for mixing and the activation energy needed to start the transesterification reaction and it increases the chemical reaction speed, the efficient molar ratio of methanol to oil, the yield of transesterification of vegetable oils and animal fats into biodiesel with less energy consumption than for conventional mechanical stirring methods.[159–161]

Transesterification is also carried out using high-frequency microwave irradiation. Microwave irradiation increases the rate of chemical reactions, reducing the time from hours to minutes and minutes to seconds. Microwave radiations mainly consist of infrared and radio waves. Generally the wavelength of microwaves lies between 1 mm and 1 m and frequency from 300 MHz to 300 GHz.[162] Used cooking oil was transesterified using 20% methanol and 1% NaOH at 65 °C.[163] The reaction was completed in 60 min and the separation phase was completed in 8 h using a conventional method. The process was repeated with same amounts of alcohol and catalyst using microwave irradiation. The reaction was completed within 2 min and the separation phase was completed in 30 min. The conversion efficiency was 100%, as compared to 96% for the conventional method. There are also various reports of the use of this technique for biodiesel production.[124,164–169]

2.7.5 *In situ* Biodiesel Techniques

The *in situ* biodiesel production is a novel technique to convert oil to biodiesel which was developed by Harrington and D'Arcy Evans in 1985.[170] In this technique, the oilseeds are directly treated at ambient temperature and pressure with a methanol solution in which the catalyst has been previously dissolved. It implies that the oil in the oilseeds is not isolated prior to transesterification to fatty acid esters.[170–172] To reduce the alcohol requirement for high efficiency during *in situ* transesterification, the oilseeds need to be dried before the reaction takes place.[173] Milled oilseeds are then mixed with alcohol in which the catalyst had been dissolved and the mixture is heated under reflux for 1–5 h. Two layers are formed around the time of completion of the reaction. The lower layer is the alcohol phase and can be recovered. The upper layer, including the crude biodiesel, is washed with water to remove the contaminants until the washing solution is neutral. After the washing step, the upper layer is dried over anhydrous sodium sulfate, then filtered, and the residual product is biodiesel.[174] However, this process cannot handle waste cooking oils and animal fats.

2.8 Purification of Crude Biodiesel

Once the transesterification reaction is completed, two major products exist: esters (biodiesel) and glycerol. The glycerol phase is much denser than the biodiesel phase and hence settles at the bottom of the reaction vessel, allowing it to be separated from the biodiesel phase. Phase separation can be observed within 10 min and can be completed within several hours of settling. In some cases, a centrifuge may be used to separate the two phases.[102]

As already discussed, both the biodiesel and glycerol are contaminated with unreacted catalyst, alcohol, and oil during the transesterification step. Therefore, crude biodiesel needs to be purified before use. The influence of the different parameters on biodiesel production is discussed in detail in later sections of this chapter. After separation from the glycerol phase, crude biodiesel may be contaminated with residual catalyst, water, unreacted alcohol, free glycerol, and soaps that were generated during the transesterification reaction.[175] Crude biodiesel enters a neutralization step and then passes through an alcohol stripper before the washing step. In some cases, acid is added to crude biodiesel to neutralize any remaining catalyst and to remove any soap.[23] Soaps react with the acid to form water-soluble salts and FFA. Neutralization before the washing step reduces the materials required for the washing step and minimizes the potential for emulsions being formed during the washing step. Unreacted alcohol should be removed with distillation equipment before the washing step to prevent excess alcohol from entering the wastewater effluent.[102] The primary purpose of this step is to wash out the remnants of the catalyst, soaps, salts, residual alcohol, and free glycerol from the crude biodiesel. Three main techniques are used for purifying biodiesel: water washing, dry washing, and membrane extraction.[176–178] These techniques are discussed in detail below.

2.8.1 Water Washing

Since both glycerol and alcohol are highly soluble in water, water washing is very effective for removing both contaminants. It also can remove any residual sodium salts and soaps. The primary material for water washing is distilled warm water or softened water (slightly acidic) because warm water prevents the precipitation of saturated fatty acid esters and retards the formation of emulsions with the use of a gentle washing action, and softened water eliminates calcium and magnesium contamination and neutralizes any remaining alkali catalysts.[102,179,180] After washing a number of times, the water phase becomes clear, signifying that the contaminants have been completely removed. Then, the biodiesel and water phases are separated by a separating funnel or centrifuge.[181] Due to the immiscibility of water and biodiesel, molecular sieves and silica gels *etc.*, can also be used to remove water from the biodiesel.[102] The remaining water can be removed from the biodiesel by passing the product over heated Na_2SO_4 (25 wt% of the amount of the ester product) overnight and then be removed by filtration.[182] However, there are many disadvantages of this method, like increased cost and production time, polluting liquid effluent and product loss. Moreover, emulsions can form when washing biodiesel made from waste cooking oils or acidic feedstocks because of soap formation.[105]

2.8.2 Dry Washing

The dry washing technique involves replacing the water with an ion-exchange resin or a magnesium silicate powder in order to remove impurities.[176] These two dry washing techniques can bring the free glycerol level down and reduce the soap content. Both the ion-exchange process and the magnesol process have the advantage of being water-less and thus eliminate many of the aforementioned problems.

2.8.3 Membrane Extraction

Contaminants can be removed using a hollow fiber membrane such as polysulfone.[177] A hollow fiber membrane, 1 m long and 1 mm in diameter filled with distilled water is immersed into the reactor at 20 °C. The crude biodiesel is pumped into the hollow fiber membrane (flow rate: 0.5 ml min^{-1}; operating pressure: 0.1 MPa). Following this step, biodiesel is passed over heated Na_2SO_4 and then filtered to remove any remaining water.[178] Emulsification is effectively avoided during the washing step and decreases loss during the refining process.

For commercial purposes, the finished biodiesel must be analyzed using analytical equipment to ensure it meets international standards. All of the properties of biodiesel, testing techniques and limits will be discussed in Chapter 3.

2.9   Refining Crude Glycerol and the use of Glycerol

Biodiesel is the desired product from the reaction, however the refining of glycerol is also important due to its numerous applications in different industrial products such as moisturizers, soaps, cosmetics, medicines, and other glycerol products.[183–185] Glycerol has good reactivity on sump oil, and is extremely effective for washing shearing shed floors, thus it can be used as a heavy duty detergent and degreaser. Glycerol can be fermented to produce ethanol, which means more biofuel can be produced.[186] The glycerol produced is only 50% glycerol or less in composition and mainly contains water, salts, unreacted alcohol, and unused catalyst.[102] The unused alkali catalyst is usually neutralized by an acid. In some cases, hydrochloric or sulfuric acids are added to the glycerol phase during the re-neutralization step and produce salts such as sodium chloride or potassium sulfate, the latter of which can be recovered for use as a fertilizer.[187] Generally, water and alcohol are removed to produce 80–88% pure glycerol that can be sold as crude glycerol. In more sophisticated operations, the glycerol is distilled to 99% or higher purity and sold in different markets.[188]

2.10 Influence of the Different Parameters on Biodiesel Production

There are important parameters that influence the biodiesel production process. In order to obtain the maximum yield of biodiesel, these parameters must be optimized.

2.10.1 Molar Ratio of Alcohol to Oil

The yield of alkyl ester increases when the molar ratio of oil to alcohol increases.[153] It has been reported that a 6 : 1 molar ratio during acid esterification and a 9 : 1 vegetable-oil-to-alcohol molar ratio during alkaline esterification are the optimum ratios for biodiesel production from high-FFA rubber seed oil and polanga seed oil, respectively.[106, 189]

2.10.2 Reaction Temperature

A higher reaction temperature can decrease the viscosities of oils and result in an increase in reaction rate as more energy is being supplied for the reaction to occur. Thus the yield of the biodiesel product is improved. The rate of reaction is dependent upon the reaction temperature. In the alkali transesterification reaction, the temperature maintained by researchers during different steps ranges between 318 and 338 K. The boiling point of methanol is 337.9 K. Temperatures higher than this will burn the alcohol and will result in much lower yield. It has been observed that temperatures higher than 323 K have a negative impact on the product yield for neat oil, but a positive effect for waste oil with higher viscosities.[35] Therefore, the reaction temperature must be less than the boiling point of alcohol (the boiling point of methanol is 60–70 °C at atmospheric pressure) to ensure the alcohol will not be lost through vaporization. Also, the yield of biodiesel decreases if the reaction temperature goes beyond its optimum level because a higher reaction temperature will accelerate the saponification reaction which results in a lower yield. Depending on the type of oil, the maximum yield is obtained at temperatures ranging from 60 to 80 °C.[139,190]

2.10.3 Water and Free Fatty Acid Content

Water can cause soap formation and frothing. The resulting soaps can induce an increase in viscosity, formation of gels and foams, and make the separation of glycerol difficult.[191,192] Water has a more negative effect than the presence of FFA, and hence the feedstock should be water-free.[27] It was shown that even a small amount of water (0.1%) in the transesterification reaction will decrease the ester conversion from vegetable oil.[193,194] The presence of water and FFA in raw materials resulted in soap formation and a decrease in yield of the alkyl ester, consumed catalyst and reduced the effectiveness of the catalyst.

The presence of water had a positive effect in the yield of methyl esters when methanol at room temperature was substituted by supercritical methanol.[195]

2.10.4 Catalyst Concentration

The presence of a catalyst such as calcium oxide (CaO) accelerated the methyl ester conversion from sunflower oil at 525 K and 24 MPa even if only a small amount of catalyst (0.3% of the oil) was added.[157] The transesterification step accelerates with the increase in CaO content from 0.3 to 3%. However, further enhancement of CaO content to 5% produced little increase in the yield of methyl ester.

2.10.5 Reaction Time

The conversion rate increases with reaction time.[196] For base-catalyzed transesterification, the yield of biodiesel reaches a maximum at a reaction time of 120 min or less.[34,197] Acid-catalyzed transesterification requires an even longer reaction time than the base-catalyzed reaction because base catalysts usually exhibit a higher reactivity than acid catalysts.[198] The reaction time needed during the conversion of TGs to biodiesel may range from 18 to 24 h.[199,200] However, an excessive reaction time will lead to a reduction in the product yield due to the back reaction of transesterification, causing more fatty acids to form soaps.[139]

2.11 Advantages of Biodiesel

Table 2.1 shows a comparison of modern available alternative fuels for transportation.[201] It is clear from the table that esters from vegetable oils (biodiesel) are the best substitutes for diesel because they do not demand any modification to diesel engines.

Biodiesel is renewable, energy efficient, can be used in most diesel engines with no or only minor modifications. It is made from either agricultural or recycled resources. Biodiesel is environmentally friendly because of its lower HC emissions, smoke and soot reductions, lower CO emissions and reduced production of greenhouse gases. Biodiesel contains no sulfur. Biodiesel does

Table 2.1 Comparison of modern available alternative fuels for transportation.

Type of fuel	Current availability	Future availability
Gasoline	Excellent	Moderate–poor
Compressed natural gas	Excellent	Moderate
Hydrogen fuel cell	Poor	Excellent
Biodiesel	Moderate	Excellent

not produce greenhouse effects, because the balance between the amount of CO_2 emitted and the amount of CO_2 absorbed by the plants producing the vegetable oil is equal.

Biodiesel is biodegradable and renewable and can help reduce dependency on foreign oil. It helps to lubricate the engine itself, decreasing engine wear. Biodiesel can be used directly in compression ignition engines with no substantial modifications of the engine. B20 can be used without engine modifications. The cost of biodiesel, however, is the main obstacle for commercialization of the product. The possibility of a continuous transesterification process and the recovery of high-quality glycerol as a biodiesel by-product are primary options to be considered to lower the cost of biodiesel.

References

1. Y. Ali and M. A. Hanna, *Bioresour. Technol.*, 1994, **50**, 153.
2. A. K. Babu and D. Devaradjane, SAE Paper No. 2003-01-0767, SAE InternationalWarrendale, 2003.
3. H. Kim, B. Kang, M. Kim, Y. M. Park, D. Kim, J. Lee and K. Lee, *Catal. Today*, 2004, **93–95**, 315.
4. G. P. A. G. Pousa, A. L. F. Santos and P. A. Z. Suarez, *Energy Policy*, 2007, **35**, 5393.
5. S. Kerschbaum and G. Rinke, *Fuel*, 2004, **83**, 287.
6. D. Darnoko and M. Cheryan, *J. Am. Oil Chem. Soc.*, 2000, **77**, 1263.
7. E. H. Pryde, *J. Am. Oil Chem. Soc.*, 1983, **60**, 1557.
8. S. P. Singh and D. Singh, *Renewable Sustainable Energy Rev.*, 2010, **12**, 200.
9. M. Ziejewski, G. L. Pratt and H. Goettler, SAE Paper No. 860301SAE International, Warrendale, 1986.
10. F. Karaosmonoglu, *Energy Sources*, 1999, **21**, 221.
11. A. W. Schwab, G. J. Dykstra, E. Selke, S. C. Sorenson and E. H. Pryde, *J. Am. Oil Chem. Soc.*, 1988, **65**, 1781.
12. S. Jain and M. P. Sharma, *Renewable Sustainable Energy Rev.*, 2010, **14**, 763.
13. C. E. Goering, 1984, Contract no. 59-2171-1-6-057-0, US Department of Agriculture, Peoria, USA.
14. N. O. V. Sonntag in *Bailey's Industrial Oil and Fat Products*, ed. D. Swern, John Wiley & Sons, New York, 4th edn, 1979.
15. P. B. Weisz, W. O. Haag and P. G. Rodeweld, *Science*, 1979, **206**, 57.
16. M. F. Demirbas, *Energ. Edu. Sci. Technol.*, 2008, **22**, 59.
17. A. W. Schwab, M. O. Bagby and B. Freedman, *Fuel*, 1987, **66**, 1372.
18. Y. C. Sharma, and B. Singh, *Fuel*, 2008, **87**, 1740.
19. F. Ma and M. A. Hanna, *Bioresour. Technol.*, 1999, **70**, 1.
20. E. Crabbe, C. N. Hipolito, G. Kobayashi, K. Sonomoto and A. Ishizaki, *Process Biochem.*, 2001, **37**, 65.
21. H. Fukuda, A. Kondo and H. Noda, *J. Biosci. Bioenergy*, 2001, **92**, 405.

22. M. P. Dorado, E. Ballesteros, J. M. Arnal, J. Gomez and F. J. Lopez, *Fuel*, 2003, **82**, 1311.
23. J. V. Gerpan, *Fuel Process. Technol.*, 2006, **86**, 1097.
24. J. M. Marchetti and A. F. Errazu, *Biomass Bioenergy*, 2008, **32**, 892.
25. V. K. Tyagi, and A. K. Vasishtha, *J. Am. Oil Chem. Soc.*, 1996, **73**, 499.
26. S. Saka and D. Kusdiana, *Fuel*, 2001, **80**, 225.
27. D. Kusdiana and S. Saka, *Bioresour. Technol.*, 2004, **91**, 289.
28. A. Demirbas, *Prog. Energy Combust. Sci.*, 2005, **31**, 466.
29. N. C. O. Tapanes, D. A. Gomes Aranda, J. W. de Mesquita Carneiro and O. A. Ceva Antunes, *Fuel*, 2008, **87**, 2286.
30. S. Sinha, A. K. Agarwal and S. Garg, *Energy Convers. Manage.*, 2008, **49**, 1248.
31. P. Felizardo, M. J. N. Correia, I. Raposo, F. J. Mendes, R. Berkemeier and B. Moura, *J. Waste Manage.*, 2006, **26**, 487.
32. U. Rashid, F. Anwar, B. R. Moser and S. Ashraf, *Biomass Bioenergy*, 2008, **32**, 1202.
33. N. Hoda, *Energy Sources, Part A*, 2010, **32**, 434.
34. H. J. Berchmans and S. Hirata, *Bioresour. Technol.*, 2008, **99**, 1716.
35. D. Y. C. Leung and Y. Guo, *Fuel Process. Technol.*, 2006, **87**, 883.
36. K. G. Georgogianni, M. G. Kontominas, E. Tegou, D. Avlonitis and V. Vergis, *Energy Fuels*, 2007, **21**, 3023.
37. Z. I. Z. Lubes and M. Zakaria, *Malaysian J. Biochem. Mol. Biol.*, 2009, **17**, 5.
38. M. J. Nye and P. H. Southwell, in *Proceedings from vegetable oil as diesel fuel*. Seminar III, Northern agricultural energy center, Peoria, IL, USA, Oct. 19–20, 1983, 78.
39. X. Meng, G. Chen and Y. Wang. *Fuel Process. Technol.*, 2008, **89**, 851.
40. J. M. Encinar, J. F. Gonzalez and R. A. Reinares, *Ind. Eng. Chem. Res.*, 2005, **44**, 5491.
41. A. Isigigur, F. Karaosmanoglu and H. A. Aksoy, *Appl. Biochim. Riotechiwl.*, 1994, **45/46**, 103.
42. A. V. Tomasevic and S. S. Siler-Marinkovic, *Fuel Process. Technol.*, 2003, **81**, 1.
43. K. Rodjanakid and C. Charoenphonphanich, presented at the 2nd Joint International Conference on Sustainable Energy and the Environment, Hua Hin, 2004.
44. B. Freedman, R. Butterfield and E. Pryde, *J. Am. Oil Chem. Soc.*, 1986, **63**, 1375.
45. Y. Li, X. Zhang and S. Li, *Energy Convers. Manage.*, 2010, **51**, 2307.
46. U. Rashid and F. Anwar, *Energy Fuels*, 2008, **22**, 1306.
47. A. Casas, C. M. Fernández, M. J. Ramos, Á. Pérez and J. F. Rodríguez, *Fuel*, 2010, **89**, 650.
48. J. M. Encinar, J. F. González and A. Rodríguez-Reinares, *Fuel Process. Technol.*, 2007, **88**, 513.

49. U. Rashid, F. Anwar and G. Knothe, *Fuel Process. Technol.*, 2009, **90**, 1157.
50. W. Xie, H. Peng and L. Chen, *Appl. Catal. A: Gen.*, 2006, **300**, 67.
51. G. J. Suppes, M. P. A. Dasari, E. J. Doskocil, P. J. Mankidy and M. J. Goff, *Appl. Catal. A: Gen.*, 2004, **257**, 213.
52. M. Zabeti, M. Wan and D. Wan, *Fuel Process. Technol.*, 2009, **90**, 770.
53. T. F. Dossin, M. F. Reyniers, R. J. Berger and G. B. Marin, *Appl. Catal. B: Environ.*, 2006, **67**, 136.
54. L. Wang and J. Yang, *Fuel*, 2007, **86**, 328.
55. D. E. Lopez, D. A. Bruce Jr and J. G. E. Lotero, *Appl. Catal. A: Gen.*, 2005, **295**, 97.
56. T. Tateno and T. Sasaki, *US. Pat.*, 6818026, 2004.
57. M. Di Serio, R. Tesser, M. Dimiccoli, F. Cammarota, M. Nasatasi and E. Santacesaria, *J. Mol. Catal.*, 2005, **239**, 111.
58. H. Mootabadi, B. Salamatinia, S. Bhatia and A. Z. Abdullah, *Fuel*, 2010, **89**, 1818.
59. B. Salamatinia, H. Mootabadi, S. Bhatia and A. Z. Abdullah, *Fuel Process. Technol.*, 2010, **91**, 441.
60. D. J. Vujicic, D. Comic, A. Zarubica, R. Micic and G. Boskovic, *Fuel*, 2010, **89**, 2054.
61. S. J. Yoo, H. S. Lee, V. Bambang, J. Kim, J. D. Kim and Y. W. Lee. *Bioresour. Technol.*, 2010, **101**, 8686.
62. V. B. Veljkovic, O. S. Stamenkovic, Z. B. Todorovic, M. L. Lazic and D. U. Skala, *Fuel*, 2009, **88**, 554.
63. R. Liu, X. Wang, X. Zhao and P. Feng, *Carbon*, 2008, **46**, 1664.
64. S. Gryglewicz, *Bioresour. Technol.*, 1999, **70**, 249.
65. E. S. Umdu, M. Tuncer and E. Seker, *Bioresour. Technol.*, 2009, **100**, 2828.
66. D. M. Alonso, R. Mariscal, R. Moreno-Tost, M. D. Poves and M. Granados, *Catal. Commun.*, 2007, **8**, 2074.
67. X. Liu, H. He, Y. Wang and S. Zhu, *Catal. Commun.*, 2007, **8**, 1107.
68. J. L. Shumaker, C. Crofcheck, S. A. Tackett, E. Santillan-Jimenez, T. Morgan, Y. Ji, M. Crocker and T. J. Troops, *Appl. Catal. B: Environ.*, 2008, **82**, 120.
69. Q. Shu, J. Gao, Z. Nawaz, Y. Liao, D. Wang and J. Wang, *Appl. Energy*, 2010, **87**, 2589.
70. A. M. Dehkhoda, A. H. West and N. Ellis, *Appl. Catal. A.: Gen.*, 2010, **382**, 197.
71. P. L. Boey, G. P. Maniam and S. A. Hamid, *Bioresour. Technol.*, 2009, **100**, 6362.
72. N. Viriya-empikul, P. Krasae, B. Puttasawat, B. Yoosuk, N. Chollacoop and K. Faungnawakij, *Bioresour. Technol.*, 2010, **101**, 3765.
73. R. Chakraborty, S. Bepari and A. Banerjee, *Bioresour. Technol.*, 2011, **102**, 3610.

74. J. Jitputti, B. Kitiyanan, P. Rangsunvigit, K. Bunyakiat, L. Attanatho and P. Jenvanitpanjakul, *J. Chem. Eng.*, 2006, **116**, 61.
75. X. Xiao, J. W. Tierney and I. Wender, *Appl. Catal. A: Gen.*, 1999, **183**, 209.
76. M. N. Stocker, *J. Mol. Catal.*, 1985, **29**, 371.
77. S. Nakagaki, A. Bail, V. C. dos Santos, V. H. R. de Souza, H. Vrubel, F. S. Nunes and L. P. Ramos, *Appl. Catal. A: Gen.*, 2008, **351**, 267.
78. D. E. Lopez, K. Suwannakarn, D. A. Bruce and J. G. Goodwin Jr, *J. Catal.*, 2007, **247**, 43.
79. L. L. Xu, X. Yang, X. D. Yu, Y. H. Guo and Maynurkader, *Catal. Commun.*, 2008, **9**, 1607.
80. Y. Wang, H. Wu and M. H. Zong, *Bioresour. Technol.*, 2008, **99**, 7232.
81. W. Xie, X. Huang and H. Li, *Bioresour. Technol.*, 2007, **98**, 936.
82. M. J. Ramos, A. Casas, L. Rodriguez, R. Romero and A. Perez, 2008, **346**, 79.
83. K. H. Chung, D. R. Chang and B. G. Park, *Bioresour. Technol.*, 2008, **99**, 7438.
84. J. M. Marchetti and A. F. Errazu, *Fuel*, 2008, **87**, 3477.
85. L. C. Meher, D. V. Sagar and S. N. Naik, *Renewable Sustainable Energy Rev.*, 2006, **10**, 248.
86. B. X. Peng, Q. Shu, J. F. Wang, G. R. Wang, D. Z. Wang and M. H. Han, *Process. Saf. Enviro. Protect.*, 2008, **86**, 441.
87. M. Mittelbach and B. Trathnigg, *Eur. J. Lipid Sci. Technol.*, 1990, **92**, 145.
88. H. Lepper and L. Friesenhagen, *US Pat.*, 4608202, 1986.
89. J. Hancsók, F. Kovács and M. Krár, *Petrol. Coal*, 2004, **46**, 36.
90. F. J. Sprules and D. Price, *US Pat.*, 2366494, 1950.
91. L. Lianhua, L. V. Pengmei, L. Wen, W. Zhongming and Y. Zhenhong, *Biomass Bioenergy*, 2010, **34**, 496.
92. M. Mittelbach and M. Koncar, *US Pat.*, 5849939, 1998.
93. S. Turkay and H. Civelekoglu, *J. Am. Oil Chem. Soc.*, 1991, **68**, 83.
94. Y. Zhang, X. H. Lu, Y. L. Yu and J. B. Ji, presented at the International Conference on Biomass Energy Technologies, Guangzhou, 2008.
95. K. Suwannakarn, Biodiesel production from high free fatty acid content Feedstocks, Dissertation, Clemson University, USA, 2008.
96. L. Jeromin, E. Peukert, G. Wollmann and K. G. A. A. Henkel, *Eur. Pat.*, 0192035, 1986.
97. M. R. Simone, C. R. Michele, G. R. Marcelli, L. S. J. Paulo, M. B. C. Fernanda and R. L. Elizabeth, *Appl. Catal. A: Gen.*, 2008, **349**, 198.
98. W. Lou, M. Zong, and Z. Duan, *Bioresour. Technol.*, 2008, **99**, 8752.
99. N. Ozbay, N. Oktar and N. A. Tapan, *Fuel*, 2008, **87**, 1789.
100. N. Shibasaki-Kitakawa, H. Honda, H. Kuribayashi, T. Toda, T. Fukumura and T. Yonemoto, *Bioresour. Technol.*, 2007, **98**, 416.
101. H. Li and W. Xie, *Catal. Lett.* 2006, **106**, 25.

102. J. V. Gerpen, B. Shanks, R. Pruszko, D. Clements and G. Knothe, Report from Iowa State University for National Renewable Energy Laboratory NREL/SR-510-36244, Golden, Colorado USA, 2004.

103. P. K. Sahoo, L. M. Das, M. K. G. Babu and S. N. Naik, *Fuel*, 2007, **86**, 448.

104. Y. Wang, O. Shiyi, P. Liu and Z. Zhang, *Energy Convers. Manage.*, 2007, **48**, 184.

105. M. Canakci and J. V. Gerpan, *Trans. Am. Soc. Agric. Eng.*, 2001, **44**, 1429.

106. A. S. Ramadhas, S. Jayaraj and C. Muraleedharan, *Fuel*, 2005, **84**, 335.

107. V. B. Veljkovic, S. H. Lakicevic, O. S. Stamenkovic, Z. B. Todorovic and M. L. Lazic, *Fuel*, 2006, **85**, 2671.

108. S. V. Ghadge and H. Raheman, *Biomass Bioenergy*, 2005, **28**, 601.

109. Y. Zhang, M. A. Dubé, D. D. McLean and M. Kates, *Bioresour. Technol.*, 2003, **90**, 229.

110. G. R. Peterson and W. P. Scarrah, *J. Am. Oil Chem. Soc.*, 1984, **61**, 1593.

111. K. Jacobson, R. Gopinath, L. C. Meher and A. K. Dalai, *Appl. Catal. B: Environ.*, 2008, **85**, 86.

112. A. K. M. G. Dalai and L.C. Meher, presented at the IEEE EIC Climate Change Technology Conference, Article No. 4057358, Ottawa, 2006.

113. P. Suarez, A. Z. Plentz, M. S. M. Meneghetti and C. R. Wolf, *Quim. Nova.*, 2007, **30**, 667.

114. E. Lotero, Y. Liu, D.E. Lopez, K. Suwannakarn, D. A. Bruce and J. G. Goodwin Jr, *Ind. Eng. Chem. Res.*, 2005, **44**, 5353.

115. R. M. De Almeida, L. K. Noda, N. S. Goncalves, S. M. P. Meneghetti and M. R. Meneghetti, *Appl. Catal. A: Gen.*, 2008, **347**, 100.

116. S. Furuta, H. Matsuhashi and K. Arata, *Biomass Bioenergy*, 2006, **30**, 870.

117. M. A. Jackson, I. K. Mbaraka and B. H. Shanks, *Appl. Catal. A: Gen.*, 2006, **310**, 48.

118. Y. M. Park, D. W. Lee, D. K. Kim, J. Lee and K. Y. Lee, *Catal. Today*, 2008, **131**, 238.

119. Y. M. Park, S. Chung, H. J. Eom, J. Lee and K. Lee, *Bioresour. Technol.*, 2010, **101**, 6589.

120. H. Muthu, S. V. Sathiya, T. K. Varathachary, D. K. Selvaraj, J. Nandagopal and S. Subramanian, *J. Braz. Chem. Eng.*, 2010, **27**, 601.

121. Y. Liu, E. Lotero and J. G. Goodwin Jr, *J. Catal.*, 2006, **243**, 221.

122. M. Zhu, B. He, W. Shi, F. J. D. Yaohui, J. Li and F. Zeng, *Fuel*, 2010, **89**, 2299.

123. W. Shi, B. He, J. Ding, J. Li, F. Yan and X. Liang, *Bioresour. Technol.*, 2010, **101**, 1501.

124. S. Zhang, Y. G. Zu, Y. J. Fu, M. Luo, D. Y. Zhang and T. Efferth, *Bioresour. Technol.*, 2010, **101**, 931.

125. X. Liang, G. Gong, H. Wua and J. Yang, *Fuel*, 2009, **88**, 613.

126. J. Zhang, S. Chen, R. Yang and Y. Yan, *Fuel*, 2010, **89**, 2939.

127. B. Y. Giri, K. N. Rao, B. L. A. Prabhavathi Devi, N. Lingaiah, I. Suryanarayana, R. B. N. Prasad and P. S. S. Prasad, *Catal. Commun.*, 2005, **6**, 788.

128. B. Hamad, R. O. L. de Souza, G. Sapaly, M. G. C. Rocha, P. G. P. de Oliveira, W. A. Gonzalez, E. A. Sales and N. Essayem, *Catal. Commun.*, 2008, **10**, 92.

129. S. T. Kolaczkowski, U. A. Asli and M. G. Davidson, *Catal. Today*, 2009, **147S**, S220.

130. S. Li, Y. Wang, S. Dong, Y. Chen and F. Cao, *Renew. Energy*, 2009, **34**, 1871.

131. C. Cannilla, G. Bonura, E. Rombi, F. Arenaa and F. Frusteri, *Appl. Catal. A: Gen.*, 2010, **382**, 158.

132. G. Corro, N. Tellez, E. Ayala and A. Marinez-Ayala, *Fuel*, 2010, **89**, 2815.

133. G. Perin, G. Alvaro, E. Westphal, L. H. Viana, R. G. Jacob, E. J. Lenardao and M. G. M. D'Oca, *Fuel*, 2008, **87**, 2838.

134. Y. C. Lin, W. J. Lee and H. C. Hou, *Atmos. Environ.*, 2006, **40**, 3930.

135. C. Y. Lin and H. A. Liun, *Fuel*, 2006, **85**, 298.

136. N. Katada, T. Hatanaka, M. Ota, K. Yamada, K. Okumura and M. Niwa, *Appl. Catal. A: Gen.*, 2009, **363**, 164.

137. V. S. Y. Lin and D. R. Radu, *US. Pat.*, 7122688, 2006.

138. A. Macario, G. Giordano, B. Onida, D. Cocina, A. Tagarelli and A. M. Giuffre, *Appl. Catal. A: Gen.*, 2010, **378**, 160.

139. D. Y. C. Leung, X. Wu and M. K. H. Leung, *Appl. Energy*, 2010, **87**, 1083.

140. V. D. Athawale, S. C. Rathi and M. D. Bhabhe, *Sep. Purif. Technol.*, 2000, **18**, 209.

141. G. T. Jeong and D. H. Park, *Appl. Biochem. Biotechnol.*, 2007, **148**,131.

142. Y. Liu, H. Xin and Y. Yan, *Ind. Crops Prod.*, 2009, **30**, 431.

143. S. Shah, S. Sharma and M. N. Gupta, *Energy Fuels*, 2004, **18**, 154.

144. M. M. Soumanou and U. T. Bornscheuer, *Enzyme Microbial Technol.*, 2003, **33**, 97.

145. K. Suwannakarn, E. Lotero and S. J. G. Changqing Lu Jr, *J. Catal.*, 2008, **255**, 279.

146. Y. Wang, F. Zhang, S. Xu, L. Yang, D. Li, D. G. Evans and X. Duan, *Chem. Eng. Sci.*, 2008, **63**, 4306.

147. Y. Xi and R. J. Davis, *J. Catal.*, 2008, **254**, 190.

148. N. Brun, A.B. Garcia, H. Deleuze, M. F. Achard, C. Sanchez, F. Durand, V. Oestreicher and R. Backov, *Chem. Mater.*, 2010, **22**, 4555.

149. P. Winayanuwattikun, C. Kaewpiboon, K. Piriyakananon, S. Tantong, W. Thakernkarnkit, W. Chulalaksananukul and T. Yongvanich, *Biomass Bioenergy*, 2008, **32**, 1279.

150. D. G. B. Boocock, S. K. Konar, V. Mao and H. Sidi, *Biomass Bioenergy*, 1996, **11**, 43.

151. A. Demirbas, *Biodiesel: A Realistic Fuel Alternative for Diesel Engines*, Springer, New York, 2008, p. 161.

152. I. A. Mohammed-Dabo, M. S. Ahmad, A. Hamza, K. Muazu and A. Aliyu, *J. Petroleum Technol. Alternative Fuels*, 2012, 3, 42.

153. A. Demirbas, *Energy Convers. Manage.*, 2002, **43**, 2349.

154. S. Hawash, N. Kamal, F. Zaher, O. Kenawi and G. El Diwani, *Fuel*, 2009, **88**, 597.

155. J. Z. Yin, M. Xiao and J. B. Song, *Energy Convers. Manage.*, 2008, **49**, 908.

156. W. Cao, H. Han and J. Zhang, *Fuel*, 2005, **84**, 47.

157. A. Demirbas, *Energy Edu. Sci. Technol.*, 2008, **21**, 1.

158. L. Wang, H. He, Z. Xie, J. Yang and S. Zhu, *Fuel Process. Technol.*, 2007, **88**, 477.

159. A. K. Singh, S. D. Fernando and R. Hernandez, *Energy Fuels*, 2007, **21**, 1161.

160. J. B. Ji, J. L. Wang, Y. C. Li, Y. L. Yu and Z. C. Xu, *Ultrasonics*, 2006, **44**, 411.

161. D. Kumar, G. Kumar, Poonam and C. P. Singh, *Ultrason. Sonochem.*, 2010, **17**, 839.

162. P. Lidstrom, J. Tierney, B. Wathey and J. Westman, *Tetrahedron*, 2001, **57**, 9225.

163. A. A. Refaat, S. T. E. Sheltawy and K. U. Sadek, *Int. J. Environ. Sci. Technol.*, 2008, **5**, 315.

164. N. Saifuddin and K. H. Chua, *Malaysian J. Chem.*, 2004, **6**, 77.

165. Z. Yaakob, B. H. Ong, M. N. S. Kumar and S. K. Kamarudin, *Int. J. Sustain. Energy*, 2009, **28**, 195.

166. H. Noureddini, D. Harkey and V. Medikonduru, *J. Am. Oil Chem. Soc.*, 1998, **75**, 1775.

167. K. Krisnangkura and R. Simamaharnnop, *J. Am. Oil Chem. Soc.*, 1992, **69**, 166.

168. V. Lertsathapornsuk, R. Pairintra, K. Pairintra and K. Krisnangkura, *Fuel Process. Technol.*, 2008, **89**, 1330.

169. http://www.thaiscience.info/.

170. K. J. Harrington and C. D'Arcy-Evans, *Ind. Eng. Chem. Prod. Res. Dev.*, 1985, **24**, 314.

171. M. Haas, K. Scott, W. Marmer and T. Foglia, *J. Am. Oil Chem. Soc.*, 2004, **81**, 83.

172. S. Siler-Marinkovic and A. Tomasevic, *Fuel*, 1998, **77**, 1389.

173. M. Haas and K. Scott, *J. Am. Oil Chem. Soc.*, 2007, **84**, 197.

174. J. F. Qian, F. Wang, S. Liu and Z. Yun, *Bioresour. Technol.*, 2008, **99**, 9009.

175. J. Schumacher, Agricultural Marketing Policy Paper No. 22, 2007. Montana State University, Bozeman, MT, USA.

176. B. S. Cooke, C. Abrams and B. Bertram, *WO Pat.*, 2005037969, 2005.

177. A. Gabelman and S. Hwang, *J. Membr. Sci.*, 1999, **159**, 61.

178. H. He, X. Guo and S. Zhu, *J. Am. Oil Chem. Soc.*, 2006, **83**, 457.
179. Y. C. Chen, S. L. He and J. Cheng, *Nat. Sci. Ed.*, 2007, **4**, 45.
180. A. Demirbas and H. Kara, *Energy Sources, Part A*, 2006, **28**, 619.
181. Z. J. Predojevic, *Fuel*, 2008, **87**, 3522.
182. F. Karaosmanoglu, K. B. Cigizoglu, M. Tuter and S. Ertekin, *Energy Fuels*, 1996, **10**, 890.
183. G. D. Wen, Y. P. Xu, H. J. Ma, Z. S. Xu and Z. J. Tian, *Int. J. Hydrogen Energy*, 2008, **33**, 6657.
184. G. P. da Silva, M. Mack and J. Contiero, *Biotechnol. Adv.*, 2009, **27**, 30.
185. Z. Wang, J. Zhuge, H. Fang and B. A. Prior, *Biotechnol. Adv.*, 2001, **19**, 201.
186. T. Whittington, Department of Agriculture and Food, Western Australia, 2006. Ethanol International overview of Production And Policies. http://www.agric.wa.gov.au/objtwr/imported_assets/content/sust/biofuel/worldethproductiono.pdf
187. J. Duncan, Energy Efficiency Conservation Authority, 2003. http://www.globalbioenergy.org/uploads/media/0305_Duncan_-_Cost-of-biodiesel-production.pdf
188. National Biodiesel Board, *Fuel Fact Sheets*, Jefferson City, 2007.
189. P. K. Sahoo, L. M. Das, M. K. G. Babu and S. N. Naik, *Fuel*, 2007, **86**, 448.
190. B. K. Barnwal and M. P. Sharma, *Renewable Sustainable Energy Rev.*, 2005, **9**, 363.
191. S. V. Ghadge and H. Raheman, *Biomass Bioenergy*, 2005, **28**, 601.
192. I. B. Agra, S. Warnijati and M. S. Pratama, in *Proceedings of the 2nd World Renewable Energy Congress*, ed. A. A. M. Sayigh, Pergamon Press, New York, 1992.
193. S. Romano, in *Proceedings of the International Conference on Plant and Vegetable Oils as Fuels*, ASAE Publications, Fargo, 1982, p. 106.
194. M. Canakci and J. V. Gerpan, *Am. Soc. Agri. Bio Eng.*, 1999, **42**, 1203.
195. A. Demirbas, *Energy Convers. Manage.*, 2006, **47**, 2271.
196. B. Freedman, E. H. Pryde and T. L. Mounts, *J. Am. Oil Chem. Soc.*, 1984, **61**, 1638.
197. P. K. Sahoo and L. M. Das, *Fuel*, 2009, **88**, 1588.
198. K. G. Georgogianni, A. K. Katsoulidis, P. J. Pomonis, G. Manos and M. G. Kontominas, *Fuel Process. Technol.*, 2009, **90**, 1016.
199. S. H. Shuit, K. T. Lee, A. H. Kamaruddin and S. Yusup, *Fuel*, 2010, **89**, 527.
200. N. U. Soriano Jr, R. Venditti and D. S. Argyropoulos, *Fuel*, 2009, **88**, 560.
201. A. Demirbas, *Energy Convers. Manage.*, 2009, **50**, 14.

CHAPTER 3

Biodiesel Properties and Specifications

3.1 Introduction

In order for biodiesel to be used commercially as a fuel, the finished biodiesel must be analyzed using sophisticated analytical equipment to ensure it meets international standards. The European EN 14214 and the American Society for Testing and Materials (ASTM) D-6751 standards are the most commonly used. The standards established to govern the quality of biodiesel on the market are based on a variety of factors which vary from region to region like the availability of feedstock, the characteristics of the diesel fuel standards existing in each region, the predominance of diesel engine type in the region and the emission regulations governing those engines. Europe has a much larger diesel passenger car market, while in the USA markets are mainly comprised of heavier duty diesel engines. It is therefore not surprising that there are some significant differences between the two standards.

3.2 Oxidation Stability

Biodiesel oxidation is a very complex process that is affected by a variety of factors, including the composition of the fuel itself and the conditions of storage.[1–4] Due to their chemical composition; biodiesel fuels are more sensitive to oxidative degradation than fossil diesel fuels. This is especially true for fuels with a high content of di -and higher unsaturated esters, as the methylene groups adjacent to double bonds are particularly susceptible to radical attack as the first step of fuel oxidation. The hydroperoxides formed

Biodiesel: Production and Properties
Amit Sarin
© Amit Sarin 2012
Published by the Royal Society of Chemistry, www.rsc.org

may polymerize with other free radicals to form insoluble sediments and gums, which are associated with fuel filter plugging and deposits within the injection system and the combustion chamber. With appropriate precautions, oxidation can be delayed but not completely prevented. Resistance to oxidative degradation is an increasingly important issue for the successful development and viability of alternate fuels. Biodiesel oxidation stability (OS) is discussed in detail in Chapter 4.

3.2.1   Limits and Methods

The biodiesel standard EN 14214 requires OS at 110 °C with a minimum induction time of 6 h by the Rancimat method (EN 14112) and the ASTM standard D-6751 introduced a limit of 3 h for OS by the Rancimat test.

3.3   Low-Temperature Flow Properties

Low-temperature performance is one of the most important properties for users of biodiesel and it is mainly indicated by the cloud point (CP), pour point (PP) and cold filter plugging point (CFPP) of the fuel.[5–7]

3.3.1   Cloud Point

The CP is the temperature at which a sample of the fuel starts to appear cloudy, indicating that wax crystals have begun to form which can clog the fuel lines and filters in a vehicle's fuel system.

3.3.1.1   *Limits and Methods*

CP is measured according to EN-ISO 23015, ASTM D-6751 and ASTM D-2500 using the CP apparatus. CP is mentioned in the above standards but a limit is not given, rather a report is required. This is due to the varying weather conditions in different countries.

3.3.2   Pour Point

The PP is defined as the temperature at which the fuel ceases to flow. The cessation of flow results from an increase in viscosity or from the crystallization of wax from oil.

3.3.2.1   *Limits and Methods*

PP is measured according to EN-ISO 3016, ASTM D-6751 and ASTM D-97 using the PP apparatus. In the determination of PP, the sample is cooled in a glass tube under prescribed conditions and inspected at intervals of 3 °C until it no longer moves when the surface is held vertically for 65 s; the PP is then taken as 3 °C above the temperature of cessation of flow. PP is mentioned in the above standards but a limit is not given, rather a report is required. This is due to the varying weather conditions in different countries.

3.3.3   Cold Filter Plugging Point

The CFPP is the temperature at which a fuel causes a filter to plug due to its crystallization. The sample is cooled in a glass tube under prescribed conditions and inspected at intervals of 1 °C. The temperature at which ester structures crystallize is recorded as the CFPP.

3.3.3.1   Limits and Methods

CFPP is measured according to EN 14214, EN 116 and ASTM D-6751 and ASTM D-6371 using the CFPP apparatus. CFFP is mentioned in the above standards but a limit is not given, rather a report is required. This is due to the varying weather conditions in different countries.

Just as with conventional diesel fuel, precautions must be taken to ensure satisfactory low-temperature operability of biodiesel and its blends.[2] There is no single best way to assess low-temperature performance, and the existing fuel standards do not include explicit specifications for cold flow properties—for either conventional diesel or biodiesel. However, the fuel provider is generally required to give an indication of the cold flow properties by reporting the CP of the fuel.

Solids and crystals rapidly grow and agglomerate, clogging fuel lines and filters and therefore causing major operability problems. With decreasing temperature, more solids form as the material approaches the PP, the lowest temperature at which it will still flow. Saturated fatty compounds have significantly higher melting points than unsaturated fatty compounds and in a mixture they crystallize at higher temperature than the unsaturated compounds. Thus biodiesel fuels derived from fats or oils with significant amounts of saturated fatty compounds will display higher CPs, PPs and CFPPs.

3.3.4   Low-Temperature Flow Test

There is another test method for the low-temperature flow properties of conventional diesel fuels, namely the low-temperature flow test (LTFT) which is used in North America (ASTM D-4539).[2] This method has also been used to evaluate biodiesel and its blends with Nos. 1 and 2 conventional diesel fuels.

3.3.5   Cold Soak Filterability Test

In recent years, another low-temperature operability problem has been recognized, resulting from the formation of insoluble particles upon storage at cool temperatures—though generally above the CP.[8] These insoluble particles arise from precipitation of trace-level non-fatty acid methyl ester (FAME) impurities, not from the major FAME components themselves. The two major families of impurities identified as causing such precipitation problems are saturated monoglycerides and sterol glucosides.[9-14]

3.3.5.1   Limit and Method

The Cold Soak Filterability Test is measured according to ASTM D-7501 and a limit of 360 s is required. In the test, biodiesel is chilled to 40F for 16 hours, and then warmed back up to room temperature without stirring or heating the sample. It is then passed through a filter under vacuum. The time in seconds taken for cold soaked biodiesel to pass through two 0.8 micron filters and the amount of particulate matter expressed in mg/l collected on the filter. The full 300ml sample must flow through the filter in under 360 seconds. There are no European methods or limits for the Cold Soak Filterability Test.

3.4   Kinematic Viscosity

Viscosity is a measure of resistance to flow of a liquid due to internal friction caused by one part of a fluid moving over another.[15] This is a critical property because it affects the behavior of fuel injection. A higher viscosity leads to poorer fuel atomization, can cause larger droplet sizes, poorer vaporization, a narrower injection spray angle, and greater in-cylinder penetration of the fuel spray.[16-21] A low viscosity can result in an excessive wear in injection pumps and power loss due to pump leakage whereas high viscosity may result in excessive pump resistance, filter blockage, high pressure, coarse atomization and low fuel delivery rates. It has been shown that in a light-duty, common rail injection system, higher viscosity FAMEs resulted in increased delay for the start of injection, reduced injection volume, and increased injection variability.[22] Viscosity is greatly affected by temperature.[23] The factors affecting the viscosity of biodiesel will be discussed in detail in Chapter 6.

3.4.1   Limits and Methods

Kinematic Viscosity is measured according to EN-ISO 3104 where it is limited to 3.5–5.0 $mm^2 s^{-1}$, and according to ASTM D-445 where it is limited to 1.9–6.0 $mm^2 s^{-1}$.

3.5   Sulfated Ash

Ash content describes the amount of inorganic contaminants such as abrasive solids and catalyst residues, and the concentration of soluble metal soaps contained in the fuel.[24] These compounds are oxidized during the combustion process to form ash, which can contribute to injector, fuel pump, piston and ring wear, and also to engine deposits.

3.5.1 Limits and Methods

The amount of sulfated ash is measured according to EN-ISO 3987 and ASTM D-874 and limited to 0.02% m/m.

3.6 Sulfur

The effect of sulfur content on engine wear and deposits appears to vary considerably in importance and depends largely on operating conditions.[24] Fuel sulfur can also affect emission-control system performance and various limits on sulfur have been imposed for environmental reasons. B100 is essentially sulfur-free.

3.6.1 Limits and Methods

The amount of sulfur is measured according to EN-ISO 20846/20884 where it is limited to 10.0 mg kg^{-1}, and according to ASTM D-5453/D-4294 where it is limited to 15/500 mg kg^{-1}.

3.7 Alkali and Alkaline Earth Metals

Metal ions are introduced into the biodiesel fuel during the production process or during storage. Whereas alkali metals are introduced from catalyst residues, alkaline earth metals may originate from washing with hard water. Sodium and potassium are associated with the formation of ash within the engine, calcium soaps are responsible for injection pump sticking.[25] These compounds are partially limited by the sulfated ash, however tighter controls are needed for vehicles with particulate traps.

3.7.1 Limits and Methods

The amounts of Group I metals (Na and K) are measured according to EN 14108/14109 and ASTM follows EN 14538 method where they are limited to 5 mg kg^{-1}. The amounts of Group II metals (Ca and Mg) are measured according to EN 14538 where they are limited to 5 mg kg^{-1}.

3.8 Flash Point

The flash point is defined as the lowest temperature at which a fuel gives off sufficient vapors, which when mixed with air will ignites momentarily. The flash point for biodiesel is used as the mechanism to limit the level of unreacted alcohol remaining in the finished fuel. The flash point is also of importance in connection with legal requirements and for the safety precautions involved in fuel handling and storage, and is normally specified to meet insurance and fire regulations.

The flash point of pure biodiesel is considerably higher than the prescribed limits, but can decrease rapidly with increasing residual alcohol.[25] As these two aspects are strictly correlated, the flash point can be used as an indicator of the presence of methanol in the biodiesel. Flash point is used as a regulation for categorizing the transport and storage of fuels, with different thresholds from region to region, so aligning the standards would possibly require a corresponding alignment of regulations.

3.8.1   Limits and Methods

Flash point is measured according to EN-ISO 3679 where it is limited to 120 °C, and according to ASTM D-93 where it is limited to 93 °C, or 130 °C if the methanol is not measured directly.

3.9   Cetane Number

Cetane number (CN) is widely used as a diesel fuel quality parameter related to the ignition delay time and combustion quality. The higher the CN, the better the ignition properties of the fuel.[26] It is measured by matching against the blend's two reference fuels namely *n*-cetane and α-methylnaphthalene. High CNs help ensure good cold-start properties and minimize the formation of white smoke.[25] Thus, a high CN is associated with rapid engine starting and smooth combustion. A low CN causes a deterioration in this behavior and higher exhaust gas emissions (hydrocarbons and particulates). In general, biodiesel has slightly higher CNs than fossil diesel. The CN increases with increasing length of both fatty acid chains and ester groups, while it is inversely related to the number of double bonds. The factors affecting the CN will be discussed in Chapter 6.

3.9.1   Limits and Methods

CN is measured according to EN-ISO 5165 where it is limited to 51 min, and according to ASTM D-613 where it is limited to 47 min. Cetane number (CN) is widely used as a diesel fuel quality parameter related to the ignition delay time and combustion quality. The higher the CN, the better the ignition properties of the fuel. It is measured by matching against the blend's two reference fuels namely n-cetane and a-methylnaphthalene. High CNs help ensure good cold-start properties and minimize the formation of white smoke. The values 51 and 47 are the minimum calculated values for biodiesel.

3.10 Methanol or Ethanol Content

Methanol or ethanol can cause fuel system corrosion, low lubricity and adverse affects on injectors due to their high volatilities. Methanol and ethanol are also harmful to some materials in fuel distribution and vehicle fuel systems.[25] Both methanol and ethanol affect the flash point of esters.

3.10.1 Limits and Methods

The amount of methanol or ethanol is measured according to EN 14110 and is limited to 0.20% m/m, and according to ASTM follows EN 14538 method where it is limited to 0.20% m/m (MeOH).

3.11 Copper Strip Corrosion

This test serves as a measure of possible difficulties with copper, zinc, brass or bronze parts of the fuel system. The presence of acids or sulfur-containing compounds can tarnish the copper strip, causing corrosion. A copper strip is heated to 50 °C in a fuel bath for three hours, and then compared to standard strips to determine the degree of corrosion.[25] Corrosion resulting from biodiesel might be induced by some sulfur compounds and by acids, so this parameter is correlated with acid number.

3.11.1 Limits and Methods

Copper strip corrosion is measured according to EN-ISO 2160 where the degree is limited to 1, and according to ASTM D-130 where the degree is limited to 3.

3.12 Phosphorus Content

Phosphorus in FAMEs stems from phospholipids (animal and vegetable material) and inorganic salts (used frying oil) contained in the feedstock.[25] Phosphorus can damage catalytic converters used in emission-control systems and its level must be kept low. Catalytic converters are becoming more common on diesel-powered equipment as emissions standards are tightened, so low phosphorus levels are gaining importance. Biodiesel produced from USA sources has been shown to have a low phosphorus content (below 1 ppm) so the maximum specification value of 10 ppm is not problematic. Biodiesel from other sources may or may not contain higher levels of phosphorus and this specification was added to ensure that all biodiesel, regardless of the source, has a low phosphorus content.

3.12.1 Limits and Methods

Phosphorous content is measured according to EN 14107 and ASTM D-4951 where it is limited to 10 mg kg^{-1}.

3.13 Conradson Carbon Residue

The Conradson carbon residue is defined as the amount of carbonaceous matter left after evaporation and pyrolysis of a fuel sample under specific conditions. Carbon residue gives a measure of the carbon-depositing tendencies of a fuel oil. The parameter serves as a measure for the tendency of a fuel sample to produce deposits on injector tips and inside the combustion chamber when used as an automotive fuel.[24] It is considered to be one of the most important biodiesel quality criteria, as it is linked with many other parameters (discussed in Chapter 6). For these reasons, carbon residue is limited in the biodiesel specifications.

3.13.1 Limits and Methods

Conradson carbon reside is measured according to EN-ISO 10370 where it is limited to 0.30% m/m (10% sample), and according to ASTM D-4530 where it is limited to 0.050% m/m (100% sample).

3.14 Ester Content

This parameter is an important tool for determining the presence of other substances and in some cases meeting the legal definition of biodiesel (*i.e.*, monoalkyl esters).[25] Low ester content values of pure biodiesel samples may originate from inappropriate reaction conditions or from various minor components within the original oil source. A high concentration of unsaponifiable matter such as sterols, residual alcohols, partial glycerides and unseparated glycerol can lead to values below the limit. As most of these compounds are removed during distillation of the final product, distilled alkyl esters generally display higher ester contents than undistilled ones.

3.14.1 Limit and Method

Ester content is measured according to EN 14103 where it is limited to a minimum of 96.5% m/m. There is no ASTM method or limit for ester content.

3.15 Distillation Temperature

The distillation temperature (90% recovered) is an important tool, like ester content, for determining the presence of other substances and in some cases meeting the legal definition of biodiesel (*i.e.*, monoalkyl esters).

3.15.1 Limit and Method

The distillation temperature (90% recovered) is measured according to ASTM D-1160 where it is limited to 360 °C. There is no European method or limit for distillation temperature (90% recovered).

3.16 Total Contamination

Total contamination is defined as the amount of insoluble material retained after filtration of a fuel sample under standardized conditions.[25] Therefore, the total contamination is an important quality criterion, as biodiesel with high concentration of insoluble impurities tend to cause blockage of fuel filters and injection pumps. High concentrations of soaps and sediments are mainly associated with these phenomena.

3.16.1 Limit and Method

Total contamination is measured according to EN-ISO 12662 where it is limited to 24 mg kg^{-1}. There is no ASTM method or limit for total contamination.

3.17 Water and Sediments

As discussed in Chapter 2, water can be introduced into biodiesel during the final washing step of the production process and has to be reduced by drying. As biodiesel is hygroscopic, it can absorb water in concentrations up to 1000 ppm during storage. Once the solubility limit is exceeded (at about 1500 ppm of water in fuels containing 0.2% methanol), water separates inside the storage tank and collects at the bottom.[24] Appreciable amounts of water and sediment in a fuel oil tend to cause fouling of the fuel-handling facilities and cause trouble for the fuel system of a burner or engine. An accumulation of sediment in storage tanks and on filter screens can obstruct the flow of oil from the tank to the combustor. Water in middle distillate fuels can cause corrosion of tanks and equipment, and if detergent is present, the water can cause emulsions or a hazy appearance. Water is necessary to support microbiological growth at fuel–water interfaces in fuel systems. Moreover, high water contents are also associated with hydrolysis reactions, partly converting biodiesel to free fatty acids, and are linked to fuel filter blocking. Lower water concentrations, which pose no difficulties in pure biodiesel fuels, may become problematic in blends with fossil diesel, as here phase separation is likely to occur. For these reasons, maximum water content is contained in the standard specifications.

3.17.1 Limits and Methods

Water content and sediment is measured according to ASTM D-2709 and is limited to 0.050% by volume. Water content is measured according to EN-ISO 12937 and is limited to 500 mg kg^{-1}.

3.18   Acid Number

Acid number or neutralization number is a measure of the amount of free fatty acids contained in a fresh fuel sample and of free fatty acids and acids from degradation in aged samples. This test is used to determine the acidic constituents in the biodiesel. If mineral acids are used in the production process, their presence as acids in the finished fuel is also measured with the acid number. It is expressed in mg KOH required to neutralize 1 g of FAME. The sample is dissolved in a mixture of toluene and propan-2-ol containing a small amount of water and titrated potentiometrically with alcoholic KOH. The end points are noted and readings of the volumes of titrating solution are taken and used in a formula to calculate the total acid number of the biodiesel sample. As discussed in Chapter 2, acidity can on the other hand be generated during the production process. The parameter characterizes the degree of fuel ageing during storage, as it increases gradually due to degradation of the biodiesel. High fuel acidity has been discussed in the context of corrosion and the formation of deposits within the engine which is why it is limited in biodiesel specifications.

3.18.1   Limits and Methods

Acid number is measured according to EN 14104 and ASTM D-664 where it is limited to 0.5 mg KOH per g.

3.19   Free Glycerol

The content of free glycerol in the biodiesel is dependent on the production process, and high values may stem from insufficient separation or washing of the ester product.[25] The glycerol may separate in storage once its solvent (methanol) has evaporated. Free glycerol separates from the biodiesel and falls to the bottom of the storage or vehicle fuel tank, attracting other polar components such as water, monoglycerides and soaps. These can lodge in the vehicle fuel filter and can result in damage to the vehicle fuel-injection system.[24] High free glycerol levels can also cause injector coking. For these reasons free glycerol is limited in the specifications.

3.19.1   Limits and Methods

Free glycerol is measured according to EN 14105/14106 and ASTM D-6584 where it is limited to 0.02% m/m.

3.20   Total Glycerin

The total glycerin method is used to determine the level of glycerin in the fuel and includes the free glycerin and the glycerin portion of any unreacted or partially reacted oil or fat.[25] Low levels of total glycerin ensure the high conversion of the oil or fat into its monoalkyl esters has taken place. High levels of mono-, di-, and triglycerides can cause injector deposits and may adversely affect cold weather operation and filter plugging.

3.20.1 Limits and Methods

Total glycerin is measured according to EN 14105 where it is limited to 0.25% m/m, and according to ASTM D-6584 where it is limited to 0.24% m/m.

3.21 Mono-, Di- and Tri-Glycerides

In common with the concentration of free glycerol, the amount of glycerides depends on the production process. Fuels with high glyceride contents are prone to deposit formation on injection nozzles, pistons and valves.

3.21.1 Limits and Methods

Monoglycerides are measured according to EN-ISO 14105 where they are limited to 0.80% m/m. Diglycerides and triglycerides are also measured according to EN-ISO 14105 where they are limited to 0.20% m/m. There are no ASTM methods or limits for mono-, di- or tri-glycerides.

3.22 Density

The densities of biodiesels are generally higher than those of fossil diesel fuel.[8] The values depend on their fatty acid composition as well as on their purity. Density increases with decreasing chain length and increasing number of double bonds, or can be decreased by the presence of low-density contaminants such as methanol.[27]

3.22.1 Limit and Method

Density is measured according to EN-ISO 3675/12185 where it is limited to 860–900 kg m^{-3}. There is no ASTM method or limit for biodiesel density.

3.23 Iodine Value

The iodine value (IV) is a measure of the total unsaturation within a mixture of fatty acids, and is expressed as the g of iodine which will react with 100 g of biodiesel. The IV of a vegetable oil or animal fat is almost identical to that of the corresponding methyl ester.[2] The IV is determined by measuring the amount of I_2 that reacts by addition to C=C bonds; thus, the IV is directly related to FAME unsaturation. However, the IV is simply a measure of total unsaturation, while oxidative stability is more strongly affected by the number of FAME molecules having multiple double bonds. For this reason, there is some controversy about the need for an IV standard at all, and certainly about the rather restrictive maximum IV value of 120 g I_2 per 100 g biodiesel set by EN 14214. The Worldwide Fuel Charter—established by a collection of USA, European, and Japanese automobile manufacturer associations—also recommends an IV specification, but with a less restrictive allowable maximum of 130 g I_2 per 100 g biodiesel.[28] The EN limit of 120 g I_2 per 100 g excludes several promising oil sources such as soybean or sunflower seed oil, as well as grape seed oil, from serving as raw materials for biodiesel production.[29] Soybean-derived biodiesel is likely to fail the EN IV specification, but would more easily satisfy the higher IV recommendation of the Worldwide Fuel Charter. The ASTM biodiesel standard does not include an IV specification, concluding that oxidative stability is better analyzed by the Rancimat oxidative stability test (EN 14112). Others have argued that there is no need for an IV specification because the cetane number specification effectively limits unsaturation.[30]

3.23.1   Limit and Method

IV number is measured according to EN-ISO 14111 where it is limited to 120 g per 100 g. As mentioned above, there is no ASTM method or limit for IV.

3.24   Linolenic Acid Methyl Ester and Polyunsaturation Content

The limitation of unsaturated fatty acids is necessary due to the fact that heating higher unsaturated fatty acids results in polymerization of glycerides. This can lead to the formation of deposits or to deterioration of lubrication properties.[24] Three or more-fold unsaturated esters only constitute a minor share in the fatty acid pattern of various promising seed oils, which have been excluded as feedstocks according to some regional standards due to their high IV. Some biodiesel experts have suggested limiting the content of linolenic acid methyl esters and polyunsaturated biodiesel rather than the total degree of unsaturation as expressed by the iodine value. Soybean, sunflower and grape seed oils meet the linolenic acid methyl ester content (C18 : 3) imposed by the standard UNE-EN 14214, however the content of linoleic acid methyl ester, with two double bonds in the carbon chain, was very high. Therefore, a higher degree of unsaturation than 137 meant that the oils did not meet the EN standard for IV. In Spain, the maximum degree of unsaturation is 160.[31] Palm oil, rich in esters of saturated fatty acids such as palmitic (C16 : 0) and stearic (C18 : 0) acids, has a lower IV.

Table 3.1 ASTM D-6751 and EN 14214 biodiesel methods and limits.

Property/units	ASTM test method	ASTM limits	EN test method	EN limits
Oxidation stability at 110 °C/h	EN 14112	Min. 3 h	EN-ISO 14112	Min. 6 h
Cloud point/°C	D-2500	—	EN-ISO 23015	—
Pour point/°C	D-97	—	EN-ISO 3016	—
Cold filter plugging point/°C	D-6371	—	EN 116	Variable
Cold soak filterability/s (max)	D-7501	360	—	—
Viscosity at 40 °C/cSt	D-445	1.9–6.0	EN-ISO 3104	3.5–5.0
Sulfated ash/% mass	D-874	0.02 (max)	EN ISO 3987	0.02 (max)
Sulfur/% mass	D-5453 /D-4294	0.0015 (S15, max) 0.05 (S500, max)	EN ISO 20846/ 20884	0.0010 (max)
Sodium and potassium/mg kg^{-1}	—	—	EN 14108/14109	5 (max)
Calcium and magnesium/mg kg^{-1}	—	—	EN 14538	5 (max)
Flash point/°C	D-93	130 (min)	EN-ISO 3679	120 (min)
Cetane number	D-613	47 (min)	EN-ISO 5165	51 (min)
Methanol or ethanol/% mass	—	—	EN 14110	0.20%
Copper strip corrosion	D-130	3 (max)	EN-ISO 2160	1 (max)
Phosphorus/%mass	D-4951	0.001 (max)	EN 14107	0.001 (max)
Conradson carbon residue (100%)/% mass	D-4530	Max. 0.05	EN ISO 10370	0.3 (max)
Ester content/% mass	—	—	EN 14103	96.5 (min)
Distillation temperature/°C	D-1160	90% at 360 °C	—	—
Total contamination/ mg kg^{-1}	—	—	EN-ISO 12662	24 (max)
Water and sediment/% vol.	D-2709	0.05 (max)	—	—
Neutralization value/mg, KOH per g	D-664	0.5 (max)	EN-ISO 14104	0.5 (max)
Free glycerin/% mass	D-6584	0.02 (max)	EN-ISO 14105/14106	0.02 (max)
Total glycerin/% mass	D-6584	0.24 (max)	EN-ISO 14105	0.25 (max)
Monoglyceride content/% mass	—	—	EN ISO 14105	Max. 0.8
Diglyceride content/% mass	—	—	EN ISO 14105	Max. 0.2
Triglyceride content/% mass	—	—	EN ISO 14105	Max. 0.2
Density/kg m^{-3}	—	—	EN 3675	860–900
Lubricity @ 60 °C, WSD/μm	—	—	—	—

Table 3.2 Biodiesel methods and limits for Argentina, Australia, Austria, Brazil, China, Germany and India.

Property (units)	Argentina (Standards: Resolution 1283/2006)	Australia (Standards: Department of Environment, Water, Heritage and the Arts)	Austria (Standards: ON C1191)	Brazil (Standards: B100-ANP Resolution 42/2004)	China (Standards: GB/T 20828)	Germany (Standards: DIN V 51606)	India (Standards: IS 15607)
Oxidation stability at 110 °C/h (min)	6	6	—	6	6	—	6
Cloud point/°C	—	—	—	—	—	—	—
Pour point/°C	—	—	—	—	—	—	—
CFPP/°C (max)	—	Report	≤0	—	Report	—	—
Cold soak filterability/s (max)	—	—	—	—	—	—	360
Viscosity at 40 °C/cSt	3.5–5.0	3.5–5.0	3.5–5.0	—	1.9–6.0	3.5–5.0	2.5–6.0
Sulfated ash/% mass (max)	—	0.02	0.02	0.02	0.02	0.03	0.02
Sulfur/ppm (max)	10	10	100	10	50	10	50
Sodium and potassium/mg kg^{-1} (max)	—	5.0	—	10	—	—	5
Calcium and magnesium/mg kg^{-1} (max)	—	5.0	—	—	—	—	5
Flash point/°C (min)	100	120	100	100	130	110	120
Cetane number (min)	45	51	49	45	49	49	51
Methanol or ethanol/% mass	—	—	0.2	—	—	0.3	0.2
Copper strip corrosion	Class 1	Class 1	—	No. 1	No. 1	No. 1	No. 1
Phosphorus/ppm (max)	10	10	20	10	—	10	10
Conradson carbon residue (100%)/% mass (max)	—	0.05	0.05	0.05	0.3	0.05	0.05 (max)
Ester content/% mass (min)	96.5	96.5	—	—	—	—	96.5

Table 3.2 (*Continued*)

Property (units)	Argentina (Standards: Resolution 1283/2006)	Australia (Standards: Department of Environment, Water, Heritage and the Arts)	Austria (Standards: GN C1191)	Brazil (Standards: B100-ANP Resolution 42/2004)	China (Standards: GB/T 20828)	Germany (Standards: DIN V 51606)	India (Standards: IS 15607)
Distillation temperature/°C	—	360	—	360 (T-95)	—	—	90% at 360 °C
Total contamination/mg kg^{-1}	—	—	—	—	—	20	—
Water and sediment/% vol. (max)	0.05	0.05	—	0.02	—	0.03	0.05
Neutralization value/mg KOH per g (max)	0.5	0.8	0.3	0.8	0.8	0.5	0.5
Free glycerin/% mass (max)	0.02	0.02	0.02	0.02	0.02	0.02	0.02
Total glycerin/%mass	0.24	0.25	0.24	0.38	0.24	0.25	0.25
Monoglyceride content/% mass	—	—	—	1.0	—	0.8	—
Diglyceride content/%mass	—	—	—	0.25	—	0.4	—
Triglyceride content/% mass	—	—	—	0.25	—	0.4	—
Density/kg m^{-3}	875–900	860–890	850–890	Report at 20 °C	820–900	875–900	860–900

Table 3.3 Biodiesel methods and limits for Indonesia, Japan, New Zealand, Philippines, South Africa, Taiwan and Worldwide Fuel Charter.

Property (units)	Indonesia (Standards: SNI 04-7182-2006)	Japan (Standards: JIS K2390)	New Zealand (Standards: Regulation SR 2008/138)	Philippines (Standards: DPNS/DOE QS 002 : 2007)	South Africa (Standards: SANS 1935 : 2004)	Taiwan (Standards: CNS 15072K5155)	Worldwide Fuel Charter (Standards: Biodiesel Guidelines 2009)
Oxidation stability at 110 °C/h (min)	—	2	6	6	6	6	10
Cloud point/°C	18	Report	—	Report	—	—	—
Pour point/°C	—	—	—	—	—	—	—
CFPP/°C (max)	—	—	—	—	—	0	—
Cold soak filterability/s (max)	—	—	—	—	—	—	—
Viscosity at 40 °C/cSt	2.3–6.0	3.5–5.0	2.0–5.0	2.0–4.5	3.5–5.0	3.5–5.0	2.0–5.0
Sulfated ash/% mass (max)	0.02	0.02	0.02	0.02	0.02	—	0.005
Sulfur/ppm (max)	100	10	10	500	10	10	10
Sodium and potassium/mg kg^{-1} (max)	—	5.0	5.0	5.0	10	5.0	5.0
Calcium and magnesium/mg kg^{-1} (max)	—	5.0	5.0	5.0	5.0	5.0	5.0
Flash point/°C (min)	100	120	100	100	100	120	100
Cetane number (min)	51	51	51	51	45	51	51
Methanol or ethanol/% mass	—	0.2	—	0.2	0.2	0.2	0.2
Copper strip corrosion	No. 3	No. 1	No. 1	No. 1	No. 1	No. 1	Light rusting
Phosphorus/ppm (max)	10	10	10	10	10	10	4
Conradson carbon residue (100%)/% mass (max)	0.05	0.3	0.05	0.05	0.05	0.3	0.05
Ester content/% mass (min)	96.5	96.5	96.5	—	96.5	96.5	96.5

Table 3.3 (*Continued*)

Property (units)	Indonesia (Standards: SNI 04-7182-2006)	Japan (Standards: JIS K2390)	New Zealand (Standards: Regulation SR 2008/138)	Philippines (Standards: DPNS/DOE QS 002 : 2007)	South Africa (Standards: SANS 1935 : 2004)	Taiwan (Standards: CNS 15072K5155)	Worldwide Fuel Charter (Standards: Biodiesel Guidelines 2009)
Distillation temperature/°C	360	360	—	360	360 (T-95)	—	—
Total contamination/mg kg^{-1}	—	24	24	—	24	24	24
Water and sediment/% vol. (max)	0.05	0.05	0.05	0.05	0.02	—	0.05
Neutralization value/mg KOH per g (max)	0.8	0.5	0.5	0.5	0.8	0.5	0.5
Free glycerin/% mass (max)	0.02	0.02	0.02	0.02	0.02	0.02	0.02
Total glycerin/%mass	0.24	0.25	0.25	0.24	0.38	0.25	0.24
Monoglyceride content/% mass	—	0.8	0.8	0.8	1.0	0.8	0.8
Diglyceride content/%mass	—	0.2	0.2	0.2	0.25	0.2	0.2
Triglyceride content/% mass	—	0.2	0.2	0.2	0.2	0.2	0.2
Density/kg m^{-3}	850–890	860–900	860–900	860–900	860–900	860–900	860–900

3.24.1 Limits and Method

Linolenic acid content is measured according to EN 14103 where it is limited to 12.0 mg kg^{-1}. There is no ASTM method or limit for linolenic acid content. Polyunsaturated ($\geq$ 4 double bonds) methyl ester content is limited in the EU 1 mg kg^{-1} however a measurement method is still in development. There is no ASTM method or limit for polyunsaturated ($\geq$ 4 double bonds) content.

3.25 Lubricity

Lubricity refers to the reduction of friction between solid surfaces in relative motion.[32] Two mechanisms contribute to overall lubricity: (a) hydrodynamic lubrication and (b) boundary lubrication. In hydrodynamic lubrication, a liquid layer (such as diesel fuel within a fuel injector) prevents contact between opposing surfaces. Boundary lubricants are compounds that adhere to the metallic surfaces, forming a thin, protective anti-wear layer. Boundary lubrication becomes important when the hydrodynamic lubricant has been squeezed out or otherwise removed from between the opposing surfaces. Good lubricity in diesel fuel is critical to protect fuel-injection systems. Biodiesel from all feedstocks is generally regarded as having excellent lubricity. Because of its naturally high lubricity, there is no lubricity specification for B100 within either the USA or European biodiesel standards. However, the USA standard for B6–B20 blends (ASTM D-7467) does include a lubricity specification, as does the conventional diesel fuel standard, ASTM D-975. Low biodiesel levels (often just 1–2%) typically provide satisfactory lubricity to diesel.[33,34]

All of the aforementioned biodiesel properties are summarized in Table 3.1 with their limits and methods.

3.26 Other Biodiesel Standards and Limits

Tables 3.2 and 3.3 list the biodiesel standards and limits for Argentina, Australia, Austria, Brazil, China, Germany, India, Indonesia, Japan, New Zealand, The Philippines, South Africa, Taiwan and the Worldwide Fuel Charter.[35–50] Most other countries follow European and ASTM methods and limits.

References

1. R. Sarin, M. Sharma, S. Sinharay and R. K. Malhotra, *Fuel*, 2007, **86**, 1365.
2. G. Knothe, *Fuel Process. Technol.*, 2005, **86**, 1059.
3. R. O. Dunn, *J. Am. Oil Chem. Soc.*, 2002, **79**, 915.
4. G. Knothe, *Energy Fuels*, 2008, **22**, 1358.
5. R. Dunn and M. O. Bagby, *J. Am. Oil Chem. Soc.*, 1995, **72**, 895.

6. R. O. Dunn, M. W. Shockley and M. O. Bagby, *J. Am. Oil Chem. Soc.*, 1996, **73**, 1719.
7. H. Imahara, E. Minami and S. Saka, *Fuel*, 2006, **85**, 1666.
8. S. K. Hoekman, A. Broch, C. Robbins, E. Ceniceros and M. Natarajan, *Renewable Sustainable Energy Rev.*, 2012, **16**, 143.
9. Conservation of clean air and water in Europe special task force, FE/STF-24, Report No. 9/09, CONCAWE, Brussels, 2009.
10. H. Y. Tang, R. C. De Guzman, S. O. Salley and K. Y. S. Ng, *J. Am. Oil Chem. Soc.*, 2008, **85**, 1173.
11. N. Ohshio, K. Saito, S. Kobayashi and S. Tanaka, SAE Paper No. 2008-01-2505, SAE International, Warrendale, 2008.
12. H. Y. Tang, S. O. Salley and K. Y. S. Ng, *Fuel*, 2008, **87**, 3006.
13. Imperial Oil, Report No. R497-2009, Natural Resources Canada, Ottawa, Canada, 2009.
14. Imperial Oil, Report No. R498-2009, Natural Resources Canada, Ottawa, Canada, 2009.
15. G. Knothe, in *The Biodiesel Handbook*, ed. G. Knothe, J. V. Gerpen and J. Krahl, AOCS Press, Urbana, 2005, p. 81.
16. C. Hasimoglu, M. Ciniviz, I. Ozsert, Y. Icingur, A. Parlak and M. S. Salman, *Renewable Energy*, 2008, **33**, 1709.
17. C. A. W. Allen and K. C. Watts, *Trans. ASAE*, 2000, **43**, 207.
18. H. K. Suh, H. G. Rho and C. S. Lee, presented at the ASME/IEEE Joint Rail Conference & Internal Combustion Engine Spring Technical Conference, Pueblo, Colorado, USA, 2007.
19. C. E. Ejim, B. A. Fleck and A. Amirfazli, *Fuel*, 2007, **86**, 1534.
20. R. Ochoterena, M. Larsson, S. Andersson and I. Denbratt, SAE Paper No. 2008-01-1393, SAE International, Warrendale, 2008.
21. E. Alptekin and M. Canakci, *Renewable Energy*, 2008, **33**, 2623.
22. S. A. Miers, A. L. Kastengren, E. M. El-Hannouny and D. E. Longman, presented at the ASME Internal Combustion Engine Division Fall Technical Conference, Charleston, South Carolina, USA, 2007.
23. G. Knothe and K. R. Steidley, *Fuel*, 2007, **86**, 2560.
24. M. Mittelbach, *Bioresour. Technol.*, 1996, **56**, 7.
25. Tripartite Task Force (Brazil, EU & USA), White Paper On Internationally Compatible Biofuel Standards, 2007.
26. L. C. Meher, D. Vidya Sagar and S. N. Naik, *Renewable Sustainable Energy Rev.*, 2006, **10**, 248.
27. A. A. Refaat, *Int. J. Env. Sci. Technol.*, 2009, **6**, 677.
28. Worldwide Fuel Charter, Report No., European Automobile Manufacturer's Association, 4th edn, European Automobile Manufacturers Association Avenue des Nerviens 85 B-1040 Brussels, Belgium, 2006. http://www.acea.be/images/uploads/aq/Final%20WWFC%204%20Sep%202006.pdf
29. M. Mittelbach and C. Remschmidt, in *Biodiesel: The Comprehensive Handbook*, ed. M. B. H. Boersedruck, Martin Mittelbach, Vienna, 2004.

30. M. Lapuerta, J. Rodriguez-Fernandez and E. F. de Mora, *Energy Policy*, 2009, **37**, 4337.
31. M. J. Ramos, C. M. Fernández, A. Casas, L. Rodríguez and Á. Pérez, *Bioresour. Technol.*, 2009, **100**, 261.
32. J. Bacha, J. Freel, L. Gibbs, G. Hemighaus, K. Hoekman, J. Horn, A. Gibbs, M. Ingham, L. Jossens, D. Kohler, D. Lesnini, J. McGeehan, M. Nikanjam, E. Olsen, R. Organ, B. Scott, M. Sztenderowicz, A. Tiedemann, C. Walker, J. Lind, J. Jones, D. Scott and J. Mills, Diesel Fuels Technical Review, Chevron Global Marketing, Chevron Corporation, California, USA, 2007.
33. National Renewable Energy Laboratory, Report No. TP-540-43672, NREL, 4th edn, Colorado, USA, 2009.
34. G. Knothe, presented at the SAE Powertrain & Fluid Systems Conference, Paper No. 2005-01-3672, San Antonio, TX, USA, 2005.
35. http://www.natbiogroup.com/docs/standards%20and%20methods%20with%20nb%2020090303.pdf
36. http://www.natbiogroup.com/docs/rnrf%20standards%20and%20methods%2020090303.pdf
37. http://www.dieselnet.com/standards/ar/fuel.php
38. http://www.environment.gov.au/atmosphere/fuelquality/publications/draft-standard.html
39. http://www.svlele.com/biodiesel_std.htm
40. http://www.dieselnet.com/standards/br/fuel.php http://ec.europa.eu/energy/renewables/biofuels/doc/standard/2007_white_paper_icbs.pdf
41. www.nist.gov/oiaa/LiHongmei.pdf http://www.bioenergytrade.org/downloads/tsukuba20kaochinabiodieselfeedstockmarket.pdf
42. http://www.tistr.or.th/APEC_website/Document/2nd%20APEC%20biodiesel/2nd%20APEC%2016Jul%2008/7.%20Biodiesel%20Situation%20in%20India.pdf
43. http://www.biodiesel-fuel.co.uk/biodiesel-standards/
44. http://www.biomass-asia workshop.jp/biomassws/05workshop/program/10_Priyanto.pdf
45. http://www.dieselnet.com/standards/jp/fuel_biodiesel.php
46. http://www.legislation.govt.nz/regulation/public/2008/0138/latest/DLM1325297.html?search=ts_regulation_Engine+Fuel_resel
47. http://www.doe.gov.ph/popup/dpns%20doe%20qs%20002%202007.pdf
48. http://www.satobiodiesel.co.za/lit_quality.html
49. http://www.biofuels.apec.org/pdfs/ewg_biodiesel_standards.pdf
50. http://www.cvma.ca/eng/publications/B100_Guideline_final_26Mar09.pdf

Oxidation Stability of Biodiesel

4.1 Introduction

As already discussed in Chapter 3, the quality of biodiesel is designated by several standards, such as EN 14214 and ASTM D-6751 and the oxidation stability (OS) is among the monitored parameters. Biodiesel oxidation is a very complex process that is affected by a variety of factors, including the composition of the fuel itself and the conditions of storage. Even if appropriate precautions are taken, oxidation can be delayed but not completely prevented. Resistance to oxidative degradation is an increasingly important issue for the successful development and viability of alternate fuels. This chapter deals with all the aspects of biodiesel stability, the factors affecting it and the various methods available to improve the OS of the biodiesel.

4.2 Factors Affecting Biodiesel Stability

The properties of vegetable oils depend upon their fatty acid composition. Common fatty acids with their chemical structures are shown in Table 4.1.[1,2]

The oxidation of fatty acid chains is a complex process, which proceeds by a variety of mechanisms. The issue of OS affects biodiesel primarily during extended storage. Oxidation of biodiesel is mainly due to unsaturation in the fatty acid chain and the presence of double bonds in the molecule offers a high level of reactivity with oxygen, especially when it is placed in contact with air/water. However, other factors such as temperature and the presence of air or metals can catalyze the oxidation of biodiesel.[3-11]

The primary oxidation products of double bonds are unstable allylic hydroperoxides which easily form a variety of secondary oxidation products.[12]

Biodiesel: Production and Properties
Amit Sarin
© Amit Sarin 2012
Published by the Royal Society of Chemistry, www.rsc.org

Table 4.1   Common fatty acids with their chemical structures.

Fatty acid	Structure (xx : y)[a]
Lauric	12 : 0
Myristic	14 : 0
Palmitic	16 : 0
Stearic	18 : 0
Arachidic	20 : 0
Behenic	22 : 0
Lignoceric	24 : 0
Oleic	18 : 1
Linoleic	18 : 2
Linolenic	18 : 3
Erucle	22 : 1

[a]xx indicates the number of carbons and y the number of double bonds in the fatty acid chain.

These include rearrangement to give short-chain aldehydes, acid compounds and high-molecular-weight materials. These compounds make biodiesel relatively unstable on storage and residual products of biodiesel such as insoluble gums, total acids, and aldehydes formed from degradation may cause engine problems like filter clogging, injector coking, and corrosion of metal parts (Table 4.2).[13–17] This is why the OS is an important criterion for biodiesel quality.

4.2.1   The Chemistry of Oxidation

The chemistry of oxidation is complicated because fatty acids usually occur in complex mixtures, with minor components in these mixtures catalyzing or

Table 4.2   The influence of residual biodiesel products on diesel engines.

Oxidation impact	Impact on fuel injection equipment	Influence on engine
Insoluble polymers	• Filter clogging	• Poor performance *i.e.*, stalling
	• Deposit formation inside the FIE	• No start
	• Injector coking	• Increased smoke
	• Seizure of moving parts	• Filter clogging
Soluble polymers	• Resins formation inside the FIE	• Poor performance *i.e.*, stalling
	• Form in soluble when blended with diesel	• No start
Total acid	• Corrosion of metal parts	• Poor performance *i.e.*, stalling
	• Foam from soap formation with metal ions from wear or corrosion	• Increased smoke
		• No start
		• Corrosion of metal parts

inhibiting oxidation. This observation affects biodiesel because usually significant amounts of esters of oleic, linoleic and linolenic acids as well as minor components which may affect oxidation are present.[18] The oxidation process can be classified into two steps, primary oxidation and secondary oxidation.

4.2.1.1 Primary Oxidation

Oxidation occurs by a set of reactions categorized as initiation, propagation, and termination as shown in Figure 4.1:[19]

In oxidation, first step (initiation) involves the removal of hydrogen from a carbon atom to produce a carbon free radical. If diatomic oxygen is present, the subsequent reaction to form a peroxy radical becomes extremely fast, preventing significant production of the alternative carbon-based free radical.[20,21] The peroxy free radical is not reactive compared to the carbon free radical, but is sufficiently reactive to quickly abstract a hydrogen from a carbon to form another carbon radical and a hydroperoxide (ROOH) in the second step (propagation). The new carbon free radical can then react with diatomic oxygen to continue the propagation cycle. This chain reaction terminates when two free radicals react with each other to yield stable products like aldehydes, alcohols and carbonic acids. These by-products formed during the oxidation process cause the fuel to eventually deteriorate. Finally, the oil spoils and becomes rancid very quickly.

During the initial period of oxidation, the ROOH concentration remains very low until a time interval, known as induction period (IP), has elapsed. The IP can be determined from the OS of the fatty oil or biodiesel under stressed conditions. Once the IP is reached, the ROOH level increases rapidly indicating the onset of the overall oxidation process Development of ROOH over time causes one the following behavior:

i. ROOH levels increase, and are then held at that level in a steady state;
ii. ROOH levels increase, achieve a peak level, and then decrease. No explanation is available for this behavior.

$$\textbf{Initiation:} \quad RH + I \rightarrow R^{\cdot} + IH$$

$$\textbf{Propagation:} \ R^{\cdot} + O_2 \rightarrow ROO^{\cdot}$$

$$ROO^{\cdot} + RH \rightarrow ROOH + R^{\cdot}$$

$$\textbf{Termination:} \ R^{\cdot} + R^{\cdot} \rightarrow R\text{-}R$$

$$ROO^{\cdot} + ROO^{\cdot} \rightarrow \text{Stable products}$$

Figure 4.1 Processes during oxidation, where I is the initiator.

4.2.1.2 Secondary Oxidation

The fatty oil hydroperoxides ultimately decompose to form hexenals, heptenals, propanal, pentane, and 2,4-heptadienal.[22-24] Increased acidity is always a result of the oxidation of fatty oils and biodiesel leading to the formation of shorter chain fatty acids.[3,5,25-27]

As hydroperoxides decompose, oxidation linking of fatty acid chains can occur to form species with higher molecular weights, known as *oxidation polymerization*. One of the results of polymer formation is an increase in the oil viscosity. Increased levels of polyunsaturated fatty acid chains increase oxidation polymerization in fatty oils. Safflower oil, which is high in linoleic (18 : 2) acid was found to exhibit a much greater increase in viscosity than safflower oil high in oleic acid (18 : 1) during air oxidation at 250 °C.[28] The increase in viscosity indicates the presence of higher molecular weight materials in the oils.

Relative rates of oxidation are 1 for oleates, 41 for linoleates, and 98 for linolenates.[29,30] This demonstrates that the OS decreases with increasing content of polyunsaturated methyl esters.

4.2.2 Measurement of Oxidation Stability

The OS of biodiesel was studied using the Rancimat equipment model 743 according to the EN 14112 and the Indian IS 15607 specifications (Figure 4.2). In the Rancimat method, oxidation is induced by passing a stream of air at a rate of 10 L h^{-1} through the biodiesel sample (3 g), at a constant temperature of 110 °C. The vapors released during the oxidation process, together with air, were passed into a flask containing 50 mL of demineralized water, containing an electrode for measuring conductivity. The electrode was connected to a measuring and recording device. It indicates the end of the IP when the conductivity begins to increase rapidly. This accelerated increase is caused by the dissociation of volatile carboxylic acids produced during the oxidation

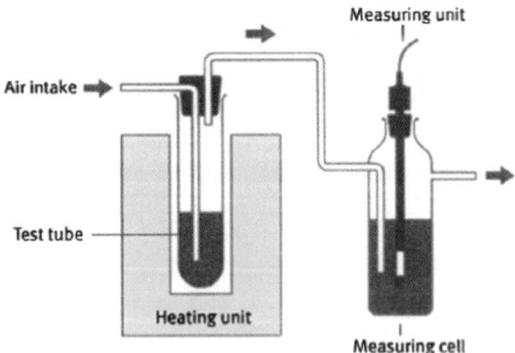

Figure 4.2   Principles of measurement for the Rancimat test method (EN 14112 and IS 15607).

process and absorbed in the water. The conductivity of this measuring solution is recorded continuously and an oxidation curve is obtained whose point of inflection is known as the IP. The biodiesel standards EN 14214 and IS 15607 require OS at 110 °C with a minimum IP of 6 h by the Rancimat method (EN 14112) and the ASTM standard D-6751 recently introduced a limit of 3 h for OS by the Rancimat test.

The OS measured using the Rancimat apparatus has been found to be similar to the oil stability index (OSI) of the American Oil Chemists' Society (AOCS) Cd 12b-93 method.[31] Under the project stability of biodiesel (BIOSTAB), samples of methyl esters from canola, sunflower oil, used frying oil and tallow were investigated to determine which method is best used to determine stability parameters.[32]

Spectroscopic methods such as nuclear magnetic resonance (NMR) and ultraviolet–visible (UV–visible) have also been used in analyzing the products of lipid oxidation.[33,34] Fourier transform infrared (FT-IR) spectroscopy has also been applied to the analysis of biodiesel degradation products resulting from accelerated oxidation in the presence of 2-ethylhexyl nitrate.[35] Based on the peaks of the carbonyl, C–O, and O–H moieties, several concomitant reaction mechanisms were proposed during oxidative degradation of biodiesel, including reverse transesterification, perester formation with ensuing secondary products, hydrolysis, and formation of various carbonyl compounds.

4.2.3 Thermal Oxidation Stability

Oxidation can also occur due to exposure of the oil to high-temperature (cooking temperature) conditions.[7,36–41] At high temperatures, the methylene-interrupted polyunsaturated olefin structure begins to isomerize to a more stable conjugated structure, then a conjugated diene group from one fatty acid chain can react with a single olefin group from another fatty acid chain to form a cyclohexene ring.[19,39,42] This reaction between a conjugated di-olefin and a mono-olefin group is known as the Diels–Alder reaction, and becomes important at temperatures of 250–300 °C or more and the products formed are called dimers.[43–45] This reaction results in the formation of carbonyl compounds such as aldehydes (formed from hydroperoxides) or high-molecular-weight polymers (formed from peroxide radicals) which increase the viscosity of biodiesel. Many authors have reported on the effect of temperature on the stability of biodiesel. The influence of the temperature on the OSI of biodiesel has been investigated and it has been reported that increasing the temperature accelerates oxidation, which decreases the OSI of the fatty acid methyl ester (FAME) contents.[7,37,40,46]

4.2.4 Storage Stability

Long-term storage tests on biodiesel have been conducted. Storage stability is the ability of liquid fuel to resist change in its physical and chemical characteristics brought about by its interaction with its environment.[47] Therefore the storage problem is accelerated by storage conditions which may include exposure to air and/or light, excess temperature and the presence of metals in the storage container of the biodiesel.[37,48] The resistance of biodiesel to oxidation degradation during storage is an important issue for the viability and sustainability of such alternative fuels.

As discussed above, the oxidation of unsaturated esters in biodiesel occurs by contact with atmospheric air and other pro-oxidizing conditions during long-term storage and impairs the fuel quality, therefore affecting the engine performance. Researchers have performed long-term storage tests on biodiesel quality and investigated the influence on the physical properties of the fuel with respect to time.[5,9,49–53] It was reported that the viscosity, density, peroxide value, and acid value of biodiesel increases but the heat of combustion decreases when it is stored for two years.[5,49,50,54] The viscosity and acid value changed dramatically over one year with variation in Rancimat IP depending on the feedstock but during 90-day storage tests, significant increases in viscosity, peroxide value, free fatty acid content, anisidine value and UV absorption were observed.[5,7,27] Biodiesel from different sources stored for 170–200 days at 20–22 °C did not exceed viscosity and acid value specifications but a decrease in the induction time with exposure to light and air, has been found to have a significant effect.[25]

The long-term storage stability of biodiesel from high oleic sunflower oil and used frying oil showed that the acid value, density and viscosity increased with increasing storage time while the iodine value decreased.[49] The researchers reported the deterioration of rapeseed oil methyl esters under different storage conditions including changes in acidity, peroxide value and viscosity, and found that the acid value, peroxide value and viscosity increased with time.[6,49,55] Polyunsaturated content has the largest impact on biodiesel stability in terms of increasing insoluble product formation and reduction in IP.[56] The storage stability of poultry fat and diesel fuel mixtures was studied with respect to specific gravity, dynamic viscosity, sedimentation accumulation and separation (layering) including corrosive effects of the fuels on various metals.[50] It was observed that the viscosity and specific gravity of these biofuels changed very little over a storage period of one year.

Long-term storage stability of biodiesel produced from karanja oil was also investigated and it was reported that the OS of karanja oil methyl ester decreased, *i.e.*, the peroxide value and viscosity increased with increasing storage time.[51] The samples stored under 'open to air inside the room' and 'exposed to metal and air' conditions had high peroxide values and viscosities compared to other conditions, hence were more susceptible to oxidation degradation.

The presence of certain metals such as Cu, Fe, Ni, Sn and brass (a copper-rich alloy) can increase the oxidizability of fatty oils.[57] The presence of Cu, even at 70 ppm in rapeseed oil greatly increased its oxidizability. Copper has also been found to reduce the OSI of methyl oleate more than either Fe or Ni.

Iron is a very effective hydroperoxide decomposer and its effect on rapeseed oil methyl esters was more pronounced at 40 °C than at 20 °C.[5]

The influence of metal contaminants on the OS of jatropha, pongamia and palm biodiesel was investigated and it was reported that even small concentrations of metal contaminants showed nearly the same influence on OS as large amounts.[9–11] Copper showed the strongest detrimental and catalytic effect on the OS. Different transition metals—Fe, Ni, Mn, Co and Cu—commonly found in metal containers were blended, as metal naphthenates, with varying concentrations (ppm) in biodiesel samples. Metal naphthenates were selected, being highly soluble in biodiesel. The metal concentration in metal naphthenates was checked using the ASTM D-4951 test method, using inductively coupled plasma atomic emission spectroscopy. The concentrations of Co, Mn, Fe, Cu and Ni in their naphthenates were 5.21, 5.20, 3.91, 6.80, and 4.99%, respectively. The samples were further diluted in biodiesel to the desired concentration. The concentration of carboxylic acids in the metal naphthenates was practically none (<1%); therefore, their significance in biodiesel will be insignificant as naphthenates were blended at the ppm level.

Figures 4.3–4.5 show that the presence of these metals depressed the OS of biodiesel, as measured by the IP. The presence of metals in biodiesel resulted in acceleration of free radical oxidation due to a metal-mediated initiation reactions.[9–11] Copper had the strongest catalytic effect and other metals—Fe, Ni, Mn and Co—also had a strong negative influence on the OS. The strong catalytic effect of Cu is due to its relative high pro-oxidant effect. Figures 4.3–4.5 shows that for all the metal contaminants, IP values became almost constant as the concentration of the metal increased. This proves that the influence of metals was catalytic, as even small concentrations of metals had nearly the same effect on the OS as large amounts. The dependence of the OS

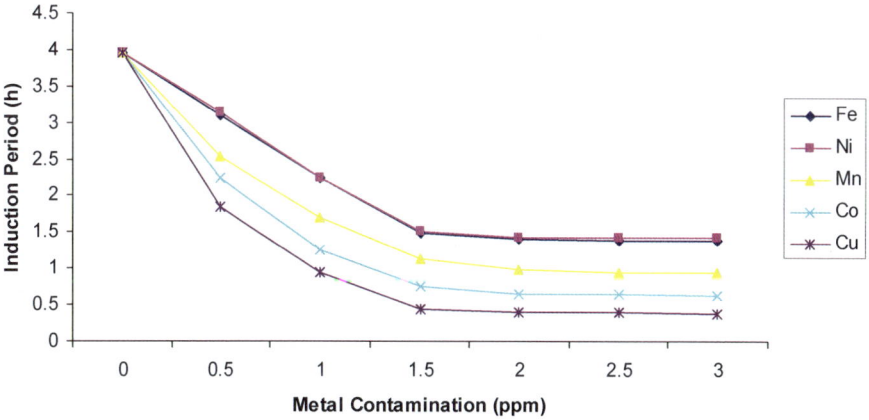

Figure 4.3 The influence of metal contamination on the OS of jatropha biodiesel.

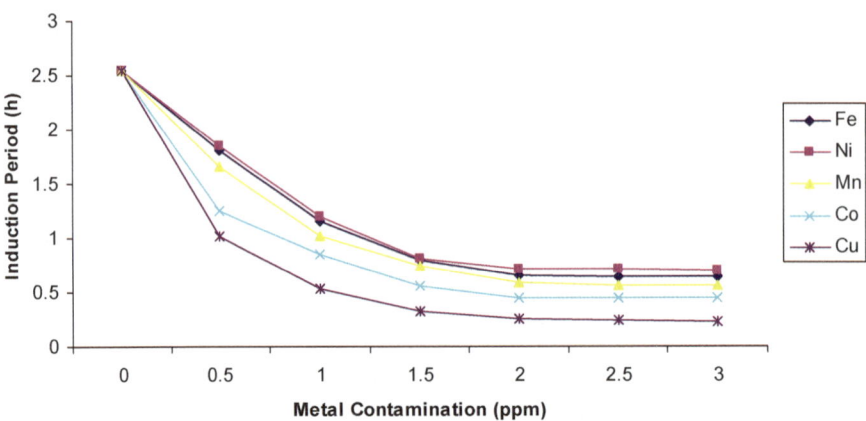

Figure 4.4   The effect of metal contamination on the OS of pongamia methyl ester.

on the type of metal confirms that the Rancimat method is a suitable lab test to correlate long-term stability.

Besides oxidation caused by exposure to air, biodiesel is also potentially subject to hydrolytic degradation due to the presence of water. This is a largely housekeeping issue although the presence of, for example, mono- and diglycerides (intermediates in the transesterification reaction) or glycerol which can emulsify water can also play a major role.[58]

It can be summarized that the nature and amount of the fatty acid chains found in biodiesel determine its OS. However, various other factors like the presence of air, light, peroxides, elevated temperatures and metals (present in the storage container of the biodiesel) can have catalytic or inhibitory effects on the OS.

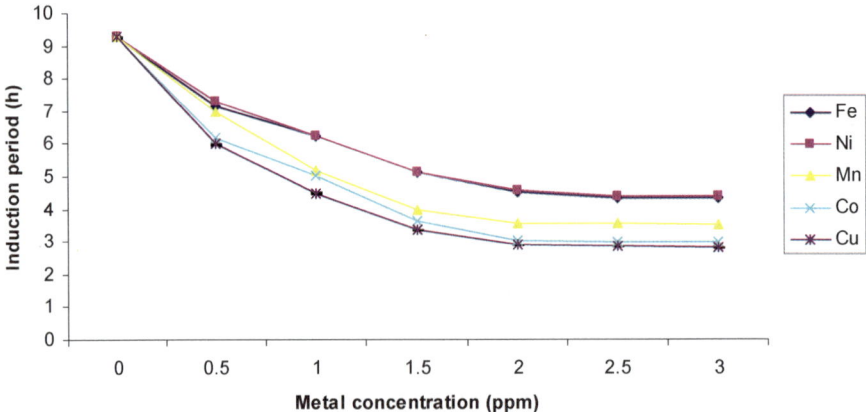

Figure 4.5   The influence of metal contamination on the OS of palm biodiesel.

4.3 Monitoring of Biodiesel Stability

4.3.1 Viscosity

Viscosity increases with the chain length (the number of carbon atoms) and with increasing degree of saturation. FFAs are responsible for higher viscosities than the corresponding methyl or ethyl esters. The double bond configuration influences the viscosity, that is a *cis* double bond configuration gives a lower viscosity than a *trans* double blond configuration while the position of the double bond affects viscosity to a lesser extent.[59] As the processes in oxidation lead to the formation of FFAs, isomerization of a double bond (usually *cis* to *trans*) and the formation of high-molecular-weight products increases the viscosity with increasing oxidation. Based on the measurement of kinematic viscosities of four different biodiesel fuels and their blends with diesel, a viscosity–temperature relationship similar to that of diesel was founded.[60]

4.3.2 Structure Indices

The allylic position equivalent (APE) and bis-allylic position equivalent (BAPE) are the theoretical measure of the number of singly allylic carbons present and the number of doubly allylic carbons present respectively in the fatty oil or ester, assuming that all poly-olefinic unsaturation is methylene-interrupted.[12] Both APE and BAPE have been correlated with OSI and peroxide value.[57] The influence of structure and concentration of individual fatty compounds on the oxidation stability of fatty acid esters was investigated. The following relations were suggested to determine the APE and BAPE values.[61]

$$APE = 2(A_{C18:1} + A_{C18:2} + A_{C18:3})$$
$$BAPE = A_{C18:2} + 2A_{C18:3}$$

where A is the amount of each fatty compound.

It has been reported that as the APE and BAPE values decrease, the IP increases indicating the credibility of the equations.

4.3.3 Iodine Value

The iodine value addresses the issue of OS because it is a measure of the total unsaturation of a fatty material measured in g of iodine per 100 g of a sample when formally adding iodine to the double bonds.[18] The iodine value also indicates the propensity of the oil or fat to polymerize and form engine deposits. The iodine value of a vegetable oil or animal fat is almost identical to that of the corresponding methyl ester, however, the iodine value of alkyl esters decreases as an increasing amount of alcohol is used in their production.

It has been suggested that the APE and BAPE are likely to be more suitable parameters than the iodine value.[61] No relationship between the iodine value and OS was observed in another investigation on biodiesel with a wide range of iodine values.[57] The iodine value is not necessarily a good method for assessing stability as this depends on the position of the double bonds available for oxidation.[61]

4.3.4 Peroxide and Acid Values

The peroxide value is less suitable for monitoring oxidation because it tends to increase and then decrease upon further oxidation due to the formation of secondary oxidation products.[4,7,26] When the peroxide value is about 350 meq per kg ester during biodiesel oxidation, the acid value and viscosity continue to increase monotonically.[4] The acid value has good potential as a parameter for monitoring biodiesel quality during storage.[15]

4.3.5 Oxidizability

Researchers have used another index of stability known as oxidizability (OX) as given by:

$$OX = [0.02(\%O) + (\%L) + 2(\%Li)]/100$$

where O refers to oleic acid (18 : 1), L refers to linoleic acid (18 : 2) and Li refers to linolenic acid (18 : 3).[62] OX decreases with increase in IP but the correlation is very low (0.2255), therefore the relationship is not a very reliable parameter.

4.4 Statistical Relationship between the Oxidation Stability and Fatty Acid Methyl Ester Composition

As discussed above, the properties of the various individual fatty esters that comprise biodiesel determine the overall fuel properties of biodiesel fuel and in turn, the properties of various fatty esters are determined by the structural features of the fatty acid that comprise a fatty ester. Researchers have also developed a statistical relationship between OS and the FAME content.[63,64]

A strong relationship was found between the OS and the content of the unsaturated fatty acid. The OS decreased with an increase of the total content of linoleic and linolenic acids. The correlation between the OS and the fatty acid content was obtained as:

$$Y = 117 : 9295/X + 2 : 5905 \ (0 < X < 100) \tag{1}$$

where X is the content of the linoleic and linolenic acids (wt%) and Y is the OS (h).

Palm biodiesel (PBD), jatropha biodiesel (JBD) and pongamia biodiesel (PoBD) were blended with different weight ratios (%) as follows: 100 : 00 : 00, 80 : 20 : 00, 80 : 00 : 20, 60 : 40 : 00, 60 : 20 : 20, 60 : 00 : 40, 40 : 60 : 00, 40 : 40 : 20, 40 : 20 : 40, 40 : 00 : 60, 20 : 80 : 00, 20 : 60 : 20, 20 : 40 : 40, 20 : 20 : 60, 20 : 00 : 80, 00 : 100 : 00, 00 : 80 : 20, 00 : 60 : 40, 00 : 40 : 60, 00 : 20 : 80, 00 : 00 : 100 respectively.[64] Their fatty acid profile was determined as shown in Table 4.3. The aforementioned 21 biodiesel samples were used during the experiment.

Figure 4.6 shows that the content of palmitic acid methyl ester (PAME) has a high degree of correlation with the OS of three biodiesel blends.[64] The correlation between OS and PAME contents (wt%) was obtained as follows:

$$OS = 0.214(PAME) + 0.671 \ (0 < PAME < 45) \tag{2}$$

For this equation, the coefficient of correlation (R) is 0.997, the coefficient of determination (R^2) is 0.994 and standard error of estimate (σ_{est}) is 0.149.

Table 4.3 Fatty acid methyl ester compositions of 12 blended biodiesel samples.

Blending ratio/wt% PBD : JBD : PoBD	Palmitic acid methyl ester/wt%	Stearic acid methyl ester/wt%	Palmitolic acid methyl ester/wt%	Oleic acid methyl ester/wt%	Linoleic acid methyl ester/wt%
00 : 00 : 00	40.3	4.1		43.4	12.2
80 : 20 : 00	35.1	4.7	0.3	43.3	16.6
80 : 00 : 20	34.4	4.4	—	49.1	12.1
60 : 40 : 00	29.8	5.2	0.6	43.3	21.1
60 : 20 : 20	28.9	5.1	0.3	49.1	16.6
60 : 00 : 40	28.1	4.9	—	54.9	12.1
40 : 60 : 00	24.6	5.8	0.8	43.2	25.6
40 : 40 : 20	23.8	5.6	0.6	49.0	21.0
40 : 20 : 40	22.9	5.5	0.3	54.9	16.4
40 : 00 : 60	22.0	5.4	—	60.7	11.9
20 : 80 : 00	19.5	6.3	1.1	43.2	29.9
20 : 60 : 20	18.6	6.2	0.8	49.0	25.4
20 : 40 : 40	17.6	6.1	0.6	54.8	20.9
20 : 20 : 60	16.8	5.9	0.3	60.6	16.4
20 : 00 : 80	15.9	5.8	—	66.4	11.9
00 : 100 : 00	14.2	6.9	1.4	43.1	34.4
00 : 80 : 20	13.3	6.8	1.1	48.9	29.9
00 : 60 : 40	12.5	6.6	0.8	54.7	25.4
00 : 40 : 60	11.5	6.5	0.6	60.6	20.8
00 : 20 : 80	10.7	6.3	0.3	66.4	16.3
00 : 00 : 100	9.8	6.2	—	72.2	11.8

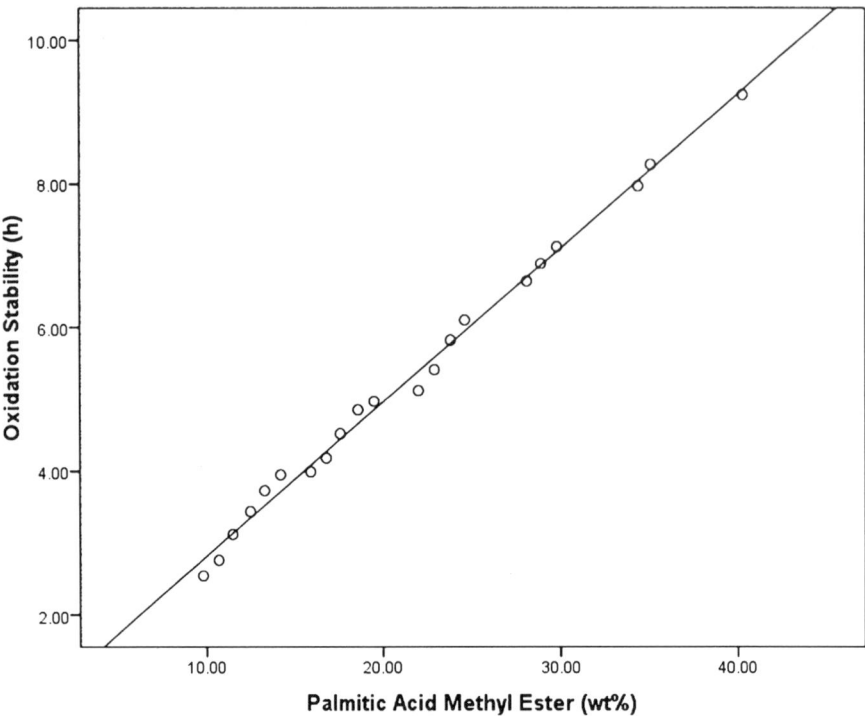

Figure 4.6   Oxidation stability of blended biodiesels with content of palmitic acid
methyl ester.

When the effects of the total unsaturated FAME content on the OS were
determined, a high degree of correlation was found (Figure 4.7).
The correlation is given as follows:

$$OS = -0.234(X) + 22.318 \ (0 < X \leq 84) \tag{3}$$

where X is the content of total unsaturated FAME (wt%). For this equation, R
= 0.999, R^2 = 0.998 and σ_{est} = 0.091.
Therefore when the compositions of blended biodiesels are known, the OS
can be predicted using the above equations. For example, when the weight
ratio of palm, jatropha, and pongamia biodiesels is 40 : 40 : 20, the content of
PAME is 23.8 wt% and the OS calculated from eqn (2) is 5.76 h which is very
close to the experimental value of 5.82 h (Figure 4.6). The minimum error of
prediction in OS values calculated from eqn (2) is 0.01 and the maximum error
is 0.24. Similarly, the OS determined from eqn (3) is 5.79 h for biodiesel blends
of weight ratios 40 : 40 : 20, when the content of total unsaturated FAME is
70.6 wt%. The minimum and maximum error of prediction in OS calculated
from eqn (8) is 0.0 and 0.22 respectively. Therefore, using eqns (2) and (3), it is

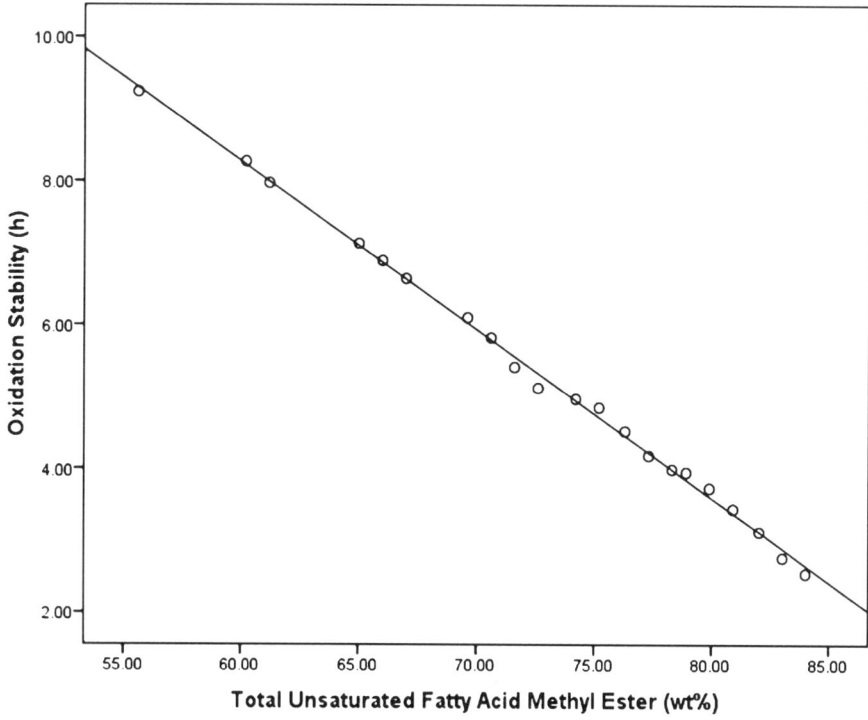

Figure 4.7 The oxidation stability of blended biodiesels with content of total unsaturated fatty acid methyl ester.

possible to directly predict the OS of biodiesel blends from the contents of PAME as well as from the total unsaturated FAME content.

4.5 How to Improve the OS of Biodiesel

4.5.1 Use of Antioxidants

Several approaches exist to either prevent oxidation as far as possible or decelerate its rate. One is obviously to prevent contact of the fatty material with air and another is to prevent contact with pro-oxidants or to avoid elevated temperatures or the presence of light. Oxidation cannot be entirely prevented but can be significantly slowed down by the use of antioxidants, which are chemicals that inhibit the oxidation process. Two types of antioxidants are generally known: chain breakers and hydroperoxide decomposers.[65] The two most common types of chain-breaking antioxidants are phenolic and amine types. Both function by donating a hydrogen atom to the peroxy radical, ROO·. The mechanism employed by all chain-breaking antioxidants is shown below.[12,65]

$$ROO^{\bullet} + AH \rightarrow ROOH + A^{\bullet}$$

$$A^{\bullet} \rightarrow \text{stable product}$$

As shown, the antioxidant contains a highly labile hydrogen that is more easily abstracted by a peroxy radical than by a fatty oil or ester hydrogen. The resulting antioxidant free radical is either stable or further reacts to form a stable molecule which is further resistant to chain oxidation processes. Therefore, chain-breaking antioxidants interrupt the oxidation chain reaction in order to enhance stability. Antioxidants may occur naturally, for example α-tocopherol (α-T), or may be synthetic and deliberately added *e.g.*, *tert*-butylated hydroxytoluene (BHT), *tert*-butylated phenol derivative (TBP), octylated butylated diphenyl amine (OBPA) and *tert*-butylhydroxquinone (TBHQ).

There are numerous publications on the effect of antioxidants on the OS of biodiesel synthesized from edible oils. Researchers synthesized phosphorylated compounds derived from cardanol, a phenolic by-product of the cashew (*Anacardium occidentale L.*) industry, and tested its application as an antioxidant for biodiesel.[66] These compounds were added to biodiesel samples in three different concentrations (500, 1000 and 2000 ppm) and their antioxidant activities were tested by thermogravimetric analysis, analyzing the onset (T_o) and endset (T_e) temperatures as reference parameters, and also by evaluating their integral procedure degradation temperatures (IPDTs). The results showed that the addition of new antioxidants increased the thermal stability of biodiesel, making this biofuel more resistant to thermo-oxidative processes.

Another study investigated the impact of various synthetic phenolic antioxidants on the OS and storage stability of PoBD.[67] The results of Rancimat experiments showed that the induction point (IP) increased substantially on adding certain antioxidants to the pongamia biodiesel. The study reveals pyrogallol (PY) to be the best antioxidant, showing the most improvement in the oxidative stability of PBD, the induction time being enhanced to 34.35 h at a PY concentration of 3000 ppm at 110 °C. The storage stability studies were carried out according to the ASTM standard procedures: (1) ASTM D-4625 at 30 °C/50 weeks and (2) ASTM D-4625 at 43 °C/12 weeks by adding different antioxidants such as BHT, BHA, PY, GA and TBHQ. The effectiveness of these five antioxidants on PBD was examined at varying loading levels during the storage period.

The efficacy of gossypol as an antioxidant additive in FAMEs prepared from soybean oil (SME), waste cooking oil (WCME) and technical grade methyl oleate (MO) was also investigated.[68] Gossypol is a naturally occurring polyphenolic aldehyde with antioxidant properties isolated from cotton seed that is toxic to humans and animals. At treatment levels of 250 and 500 ppm, gossypol exhibited statistically significant improvements in the IPs (IPs; EN 14112) of SME, WCME and MO. Efficacy was most pronounced in SME, which was due to its higher concentration of endogenous tocopherols (757 ppm) versus WCME (60 ppm) and MO (0 ppm). A comparison of antioxidant efficacy was made with BHT and γ-tocopherol (γ-T). For FAMEs with low concentrations of endogenous tocopherols (WCME and MO), γ-T exhibited the greatest efficacy, although treatments employing BHT and gossypol also yielded statistically significant improvements to oxidative stability. In summary, gossypol was effective as an exogenous antioxidant for the FAMEs investigated herein. In particular, FAMEs containing a comparatively high percentage of endogenous tocopherols were especially suited to gossypol as an antioxidant additive.

The OS of soybean oil FAME by OSI has also been studied.[69] This work examined the OSI as a parameter for monitoring the oxidative stability of soybean oil FAME. Soybean methyl ester samples from five separate sources and with varying storage and handling histories were analyzed for OSI at 60 °C using an OS instrument. Results showed that OSI may be used to measure the relative OS of SME samples as well as to differentiate between samples from different producers.

Researchers compared the oxidative stability of neutralized, refined, and frying oil waste soybean oil fatty acid ethyl ester.[70] It was observed that the biodiesel derived from neutralized, refined and, waste frying soybean oils presented different oxidative stabilities. The biodiesel from neutralized soybean oil presented a greater stability, followed by the refined and the frying waste. Due to the natural antioxidants in its composition, the neutralized soybean oil biodiesel had a higher OS.

The OS of methyl esters derived from rapeseed oil and waste frying oil, both distilled and undistilled, by differential thermal analysis and using the Rancimat method has also been studied.[46] The results obtained by both techniques were compared. Both techniques show that OS increases considerably with the addition of antioxidants. Distillation of the methyl esters prepared from rapeseed oil decreases their OS, obviously owing to the removal of natural antioxidants.

Researchers have investigated the influence of different synthetic and natural antioxidants on the OS of biodiesel produced from rapeseed oil, sunflower oil, used frying oil, and beef tallow, both distilled and undistilled.[71] Addition of the four synthetic antioxidants PY, propylgallate (PG), TBHQ, and butylated hydroxyanisole (BHA) used was found to result in an increase in the IP. These four compounds and the widely used BHT were selected for further studies at concentrations from 100 to 1000 mg kg^{-1}. The IPs of methyl esters from rapeseed oil, used frying oil and tallow were found to improve significantly with PY, PG, and TBHQ, whereas BHT was not very effective.

Researchers investigated the effectiveness of various natural and synthetic antioxidants; α-T, BHA, BHT, TBHQ, 2,5-di-*tert*-butyl-hydroquinone (DTBHQ), ionol BF200 (IB), PG, and PY to improve the oxidative stability of soybean oil (SBO), cotton seed oil (CSO), poultry fat (PF), and yellow grease (YG) based biodiesel at varying concentrations between 250 and

1000 ppm.[72] The results indicated that different types of biodiesel have different natural levels of oxidative stability, indicating that natural antioxidants play a significant role in determining oxidative stability. Moreover, PG, PY, TBHQ, BHA, BHT, DTBHQ, and IB can enhance the OS of these different types of biodiesel. Antioxidant activity increased with increasing concentration. The IP of SBO, CSO, YG, and distilled SBO-based biodiesel could be improved significantly with PY, PG and TBHQ, while PY, BHA, and BHT show the best results for PF-based biodiesels. This indicates that the effect of each antioxidant on biodiesel differs depending on the different feedstock. Moreover, the effect of antioxidants on B20 and B100 was similar; suggesting that improving the oxidative stability of biodiesel can effectively increase that of biodiesel blends. The oxidative stability of untreated SBO-based biodiesel decreased with increasing indoor and outdoor storage times, while the IP values remained constant for up to nine months when adding TBHQ to SBO-based biodiesel.

The stabilizing effect of α-, γ- and δ-tocopherols from 250 to 2000 mg kg^{-1} was evaluated using the thermal and accelerated storage induction times based on a rapid viscosity increase, in sunflower methyl ester (SuME), recycled vegetable oil methyl ester (RVOME), rapeseed methyl ester (RME) and tallow methyl ester (TME).[73] Both induction times showed that the stabilizing effect is of the order of δ- > γ- > α-tocopherol, and that the stabilizing effect increases with concentration. Tocopherols were found to stabilize methyl esters by reducing the rate of peroxide formation. The deactivation rates of tocopherols increased with unsaturation of the particular methyl ester and in the present work they were in the order SuME > RME > RVOME > TME. While α-T was found to be a relatively weak antioxidant, both γ- and δ-tocopherols increased induction times significantly and should be added to methyl esters without natural antioxidants.

The effectiveness of five antioxidants, TBHQ, BHA, BHT, PG and α-T in mixtures with soybean oil was reported and it was found that increasing the antioxidant loading (concentration) also increased the activity.[74] Results showed that PG, BHA and BHT were most effective, and α-T the least effective at loadings of up to 5000 ppm.

It has been reported that synthetic antioxidants are more effective than natural antioxidants.[75] Domingos *et al.* found that crude palm oil methyl ester containing not less than 600 ppm vitamin E exhibited an OS of more than 6 h and thus conformed to EN 14214. While distilled palm oil methyl ester needs to be treated with antioxidants in order to meet this specification. Synthetic antioxidants, namely BHT and TBHQ were found to be more effective than natural antioxidants. The efficiency of antioxidants investigated in the study was as follows: TBHQ > BHT > α-T. There are other reports of investigations into the effect of synthetic and natural antioxidants on the stability of biodiesel.[76,77]

In line with the proposed National Mission on biodiesel in India, researchers studied the oxidative stability of biodiesel synthesized from tree-borne non-edible jatropha oil seeds.[78] Neat JBD exhibited an OS of 3.95 h and research was conducted to investigate the influence of natural and synthetic antioxidants on the OS of jatropha methyl ester. Antioxidants, namely α-tocopherol, *tert*-butylated hydroxytoluene, *tert*-butylated phenol derivative, octylated butylated diphenyl amine, and *tert*-butylhydroxquinone were added to improve OS. It was found that both types of antioxidants showed beneficial effects in increasing the OS of jatropha methyl ester, but comparatively, the synthetic antioxidants were found to be more effective.

The OS of metal-contaminated biodiesel was found to increase with an increase in the dosage of antioxidant, but the dosage required for copper-contaminated biodiesel was approximately four times that required for neat biodiesel.[9–11]

The antioxidant BHT was added to metal-contaminated JBD samples at varying concentrations (ppm), and the corresponding IPs were measured using the Rancimat test method. As metallic impurities have catalytic effects, a 2 ppm metal concentration was selected for antioxidant dose optimization. Figure 4.8 shows the variation of the IP of 2 ppm metal-contaminated JBD with varying concentrations of BHT.[9]

The OS of metal-contaminated JBD has been found to increase with increasing dosage of BHT. It was found that a minimum 500 ppm dosage of BHT was needed to improve the IP of iron- and nickel-contaminated JBD and minimum dosage of 700 ppm of BHT in manganese-contaminated JBD was needed to meet the EN 14112 specification for biodiesel OS. Figure 4.8 also shows that for cobalt- and copper-contaminated JBD, minimum dosages of 900 and 1000 ppm, respectively were required to meet EN 14112 specifications. Almost the same results were found with metal-contaminated pongamia methyl ester (Figure 4.9).[10]

Antioxidants, namely BHT, TBP, OBPA and TBHQ were added to PBD samples at varying concentrations (ppm level), and the corresponding IPs were

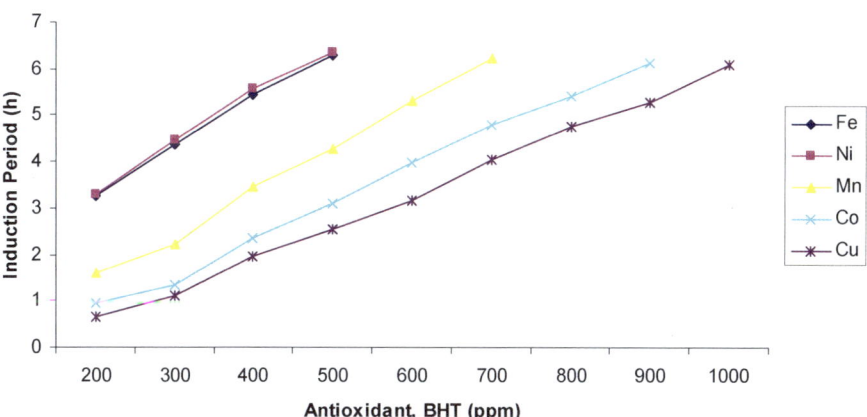

Figure 4.8 The influence of antioxidant concentration on the OS of metal-contaminated (2 ppm) jatropha methyl ester.

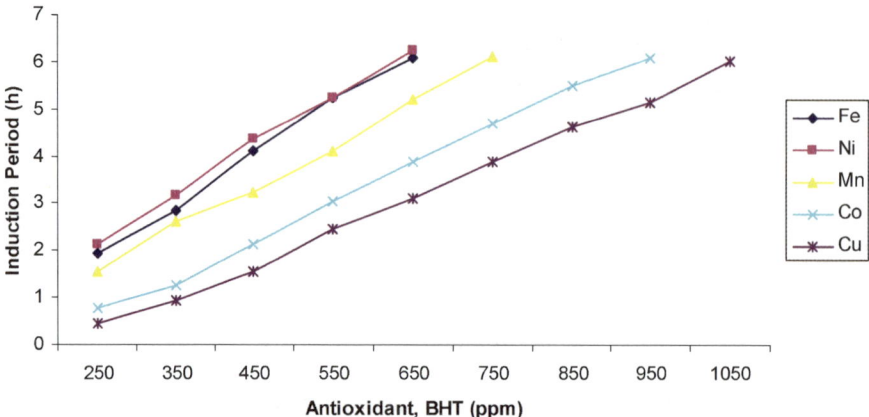

Figure 4.9   The effect of antioxidant concentration on the OS of metal-contaminated (2 ppm) pongamia methyl ester.

measured with the Rancimat test method to observe the effectiveness of different antioxidants.[11] As metallic impurities have a catalytic effect, a 2 ppm metal concentration was selected for antioxidant dose optimization.

Figure 4.10 shows the variation of the IP of PBD contaminated with 2 ppm of metal at varying concentrations of the antioxidant BHT. The OS of metal-contaminated PBD was found to increase with an increase in the dosage of the antioxidant BHT. Finally, it was found that a minimum 100 ppm dosage of BHT was needed to improve the IP of iron- and nickel-contaminated PBD and a minimum dosage of 200 ppm of BHT in manganese-contaminated PBD was needed to meet the EN 14112 specification for biodiesel OS. Figure 4.10 also shows that for cobalt- and copper-contaminated PBD, a minimum dosage of 250 ppm was required to meet EN 14112 specifications.

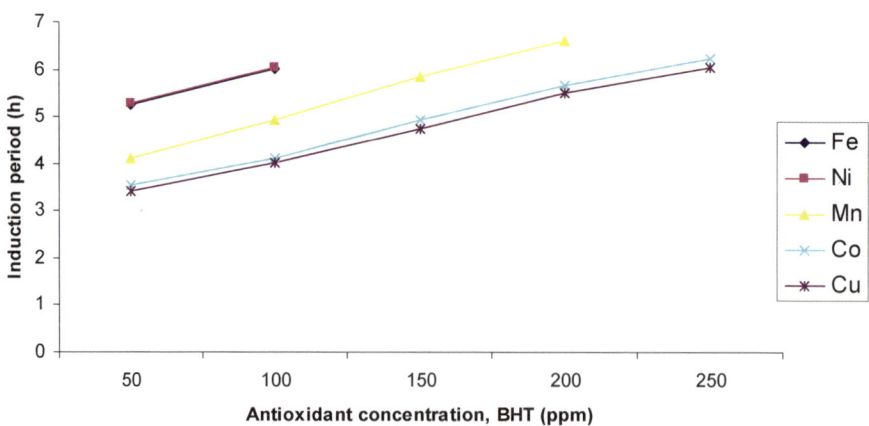

Figure 4.10   The effect of BHT concentration on the OS of metal-contaminated (2 ppm) palm biodiesel.

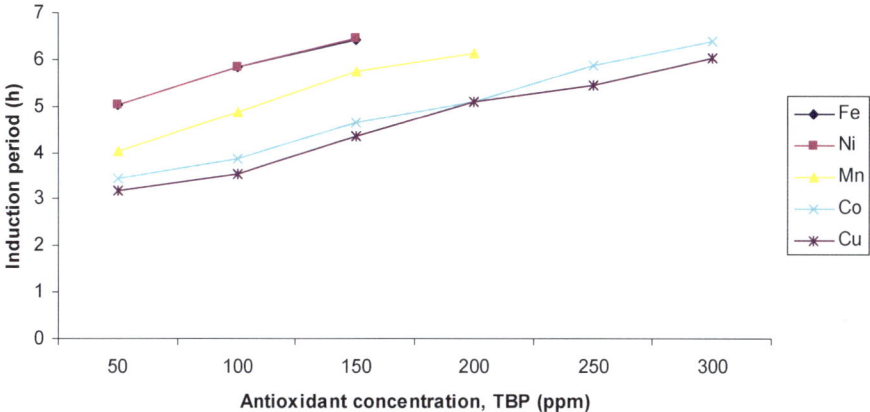

Figure 4.11   The effect of TBP concentration on the OS of metal-contaminated (2 ppm) palm biodiesel.

Figure 4.11 shows the variation of IP of PBD contaminated with 2 ppm of metal with varying concentrations of the antioxidant TBP. It was found that a minimum 150 ppm dosage of TBP was needed to improve the IP of iron- and nickel-contaminated PME and a minimum dosage of 200 ppm of TBP in manganese-contaminated PME was needed to meet the EN 14112 specification for biodiesel OS. Figure 4.11 also shows that for cobalt- and copper-contaminated PBD, a minimum dosage of 300 ppm was required to meet the EN 14112 specifications.

The antioxidant OBPA showed almost the same effect as TBP (Figure 4.12).

Further tests were done with the antioxidant TBHQ. Figure 4.13 shows the variation of the IP of PBD contaminated with 2 ppm of metal with varying concentrations of the antioxidant TBHQ. It was found that a minimum 50

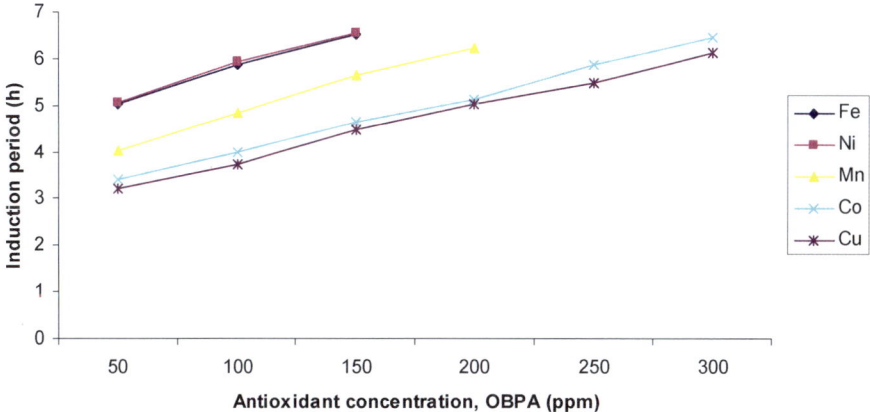

Figure 4.12   The effect of OBPA concentration on the OS of metal-contaminated (2 ppm) palm methyl ester.

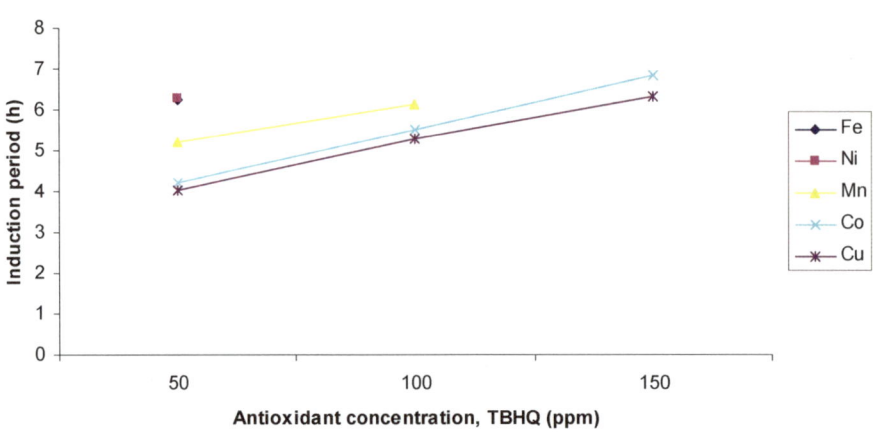

Figure 4.13 The effect of TBHQ concentration on the OS of metal-contaminated (2 ppm) palm biodiesel.

ppm dosage of TBHQ was needed to improve the IP of iron- and nickel-contaminated PBD and a minimum dosage of 100 ppm in manganese-contaminated PBD was needed to meet the EN 14112 specification for biodiesel OS. Figure 4.13 also shows that for cobalt- and copper-contaminated PBD, a minimum dosage of 150 ppm was required to meet the EN 14112 specifications. Therefore, it was found that the antioxidant TBHQ is the most effective among all the antioxidants used. This can be explained based on the molecular structures of the antioxidants. TBHQ has more OH groups attached to its aromatic ring than the others. Therefore, TBHQ offers more sites for the formation of complexes between free radicals and antioxidant radicals for lipid stabilization. Hence, TBHQ is more effective compared to other antioxidants at the same stage.

Researchers have also developed correlations between IP and different metal contaminants as shown in Table 4.4.[79] Here A is antioxidant concentration and M is the metal contaminant concentration.

Table 4.4 Correlations developed between IP and different metal contaminants.

Sample No.	Metal	Correlation	Regression coefficient
1	Fe	$IP = 0.1255(A)^{0.7334}(M)^{-0.472}$	0.93
2	Ni	$IP = 0.11735(A)^{0.7279}(M)^{-0.4663}$	0.92
3	Mn	$IP = 0.0932(A)^{0.7578}(M)^{-0.477}$	0.95
4	Co	$IP = 0.0841(A)^{0.7648}(M)^{-0.5871}$	0.95
5	Cu	$IP = 0.063(A)^{0.7986}(M)^{-0.7599}$	0.94

4.5.2 Synergistic Effects of Metal Deactivators and Antioxidants on the Oxidation Stability of Metal-Contaminated Biodiesel

Metal deactivators function by preventing the corrosion of 'yellow metals' (copper, brass and bronze) caused by the attack of acids formed during the oxidation process.[80] The metal experiences a reduction in weight due to the reaction of the acid with its surface during corrosion. There are two types of metal deactivators: surface passivators and chelators. Chelators function in the bulk of the substrate by trapping metal ions. They do not function by attaching themselves to the surface of the metal; therefore they do not eliminate the source of the metal ion. Therefore once these inhibitors are consumed, the substrate is again exposed to the ravages of metal-catalyzed oxidation.

Although it is possible to meet the desired EN specification for metal-contaminated biodiesel by using antioxidants, this has a cost implication, as antioxidants are costly chemicals. Researchers have done experiments to reduce the cost of antioxidants by adding very small amounts of the metal deactivator (MD) N,N'-disalicylidene-1,2-diaminopropane;[81] a very small amount of MD was blended with the antioxidant BHT and the synergistic effect on OS was studied in metal-contaminated JBD.

As metallic impurities have a catalytic effect, a 2 ppm metal concentration was selected for antioxidant dose optimization. Figure 4.14 shows the variation of IP of 2 ppm metal- contaminated JBD with varying concentrations of BHT. A blend of 5 ppm MD in BHT was chosen for the initial study. The OS of metal-contaminated JBD was found to increase with increasing dosage of BHT containing 5 ppm of MD. It was found that a minimum 300 ppm dosage of BHT was needed to improve the IP of iron- and nickel-

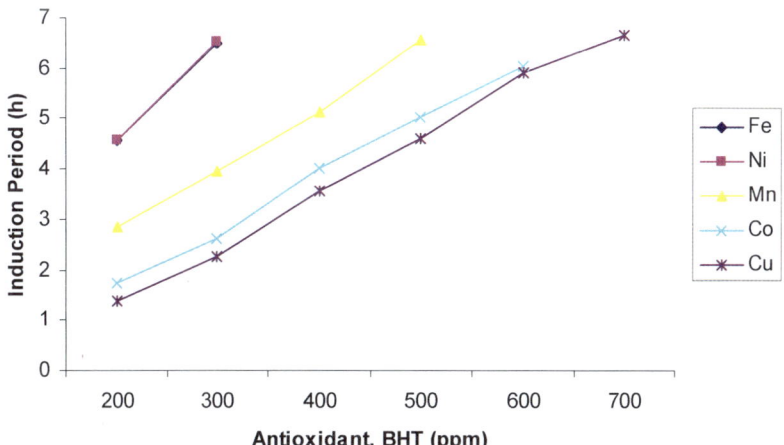

Figure 4.14   The synergistic effect of antioxidant concentration and metal deactivator (5 ppm) on the OS of metal-contaminated (2 ppm) jatropha methyl ester.

contaminated JBD and a minimum dosage of 500 ppm of BHT in manganese-contaminated JBD was needed to meet the EN 14112 specification for biodiesel OS. Figure 4.14 also shows that for cobalt- and copper-contaminated JBD, minimum dosages of 600 ppm and 700 ppm, respectively were required to meet EN 14112 specifications. Hence, it was observed that, when 5 ppm of MD was blended with BHT to enhance the IP of JBD to meet the EN 14112 specification, the minimum concentration of antioxidant required reduced to 200 ppm of BHT for iron-, nickel- and manganese-contaminated JBD, and 300 ppm of BHT for cobalt- and copper-contaminated JBD.

A concentration of 10 ppm MD was blended with BHT for a second set-up of experiments. It was observed that a minimum 250 ppm dosage of BHT was needed to improve the IP of iron- and nickel- contaminated JBD and a minimum dosage of 450 ppm of BHT in manganese-contaminated JBD was needed to meet the EN 14112 specifications for biodiesel OS (Figure 4.15). A minimum dosage of 550 ppm of BHT was required for cobalt- and copper-contaminated JBD to meet EN 14112 specifications. Hence, it was observed that, when 10 ppm of MD was blended with BHT to enhance the IP of JBD to meet the EN 14112 specification for biodiesel OS, the minimum concentration of antioxidant required reduced to 250 ppm of BHT for iron-, nickel- and manganese-contaminated JBD, 350 ppm of BHT for cobalt-contaminated JBD and 450 ppm for copper-contaminated JBD, respectively. Therefore due to the synergistic effect of antioxidant and very small amounts of MD, it can be concluded that the usage (and cost) of antioxidants can be reduced by 30–50%, to meet the EN 14112 specifications for OS.

Reduction of the usage of BHT was again noted when 15 and 20 ppm of MD was blended with BHT (Figures 4.16 and 4.17). A further increase in the concentration of MD in the antioxidant BHT showed approximately the same results.

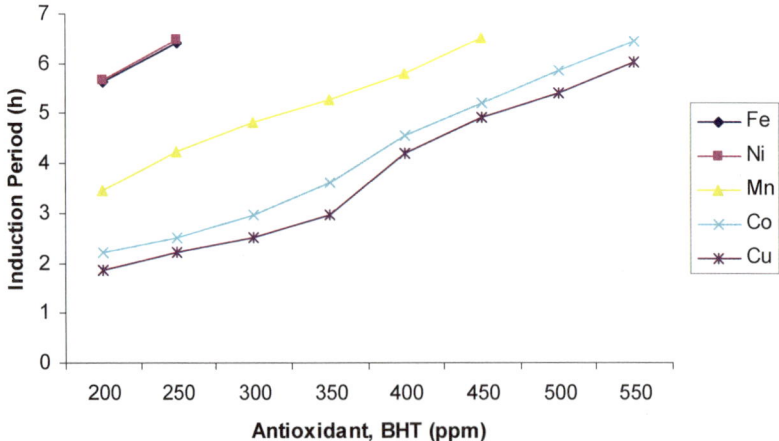

Figure 4.15   The synergistic effect of antioxidant concentration and metal deactivator (10 ppm) on the OS of metal-contaminated (2 ppm) jatropha methyl ester.

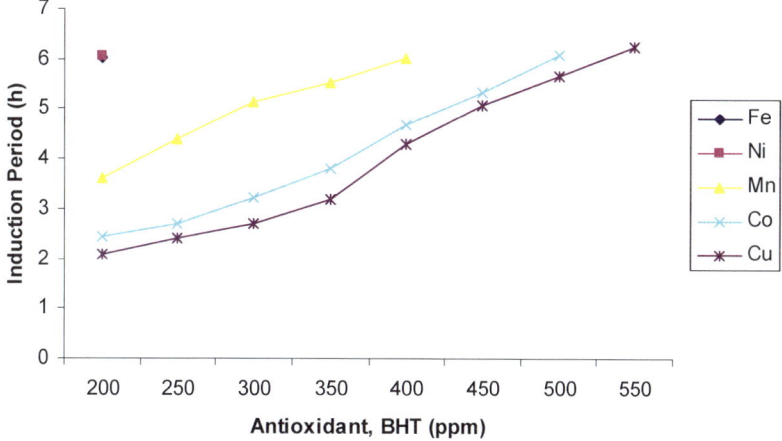

Figure 4.16 The synergistic effect of antioxidant concentration and metal deactivator (15 ppm) on the OS of metal-contaminated (2 ppm) jatropha methyl ester.

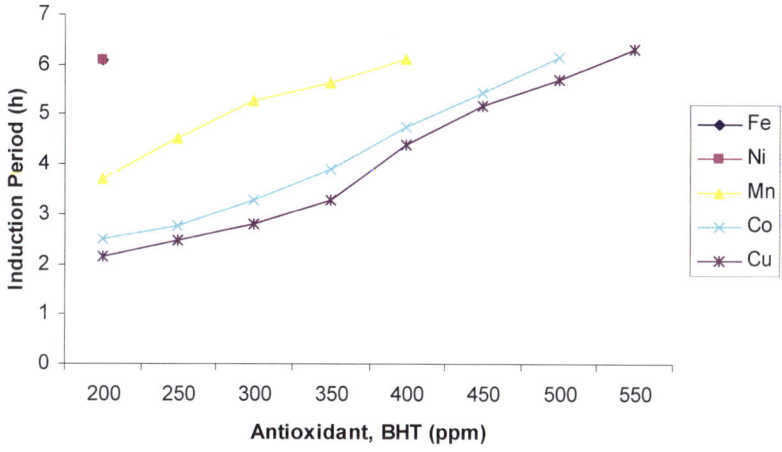

Figure 4.17 The synergistic effects of antioxidant concentration and metal deactivator (20 ppm) on the OS of metal-contaminated (2 ppm) jatropha methyl ester.

The results are comparable with previous work done on the stabilization of metal-contaminated fats and oils by combinations of metal deactivators and antioxidants.[82–86]

4.5.3 Blending of Biodiesels

Blending of biodiesels with different FAME compositions improves OS. Jatropha and pongamia biodiesels synthesized from their respective non-edible

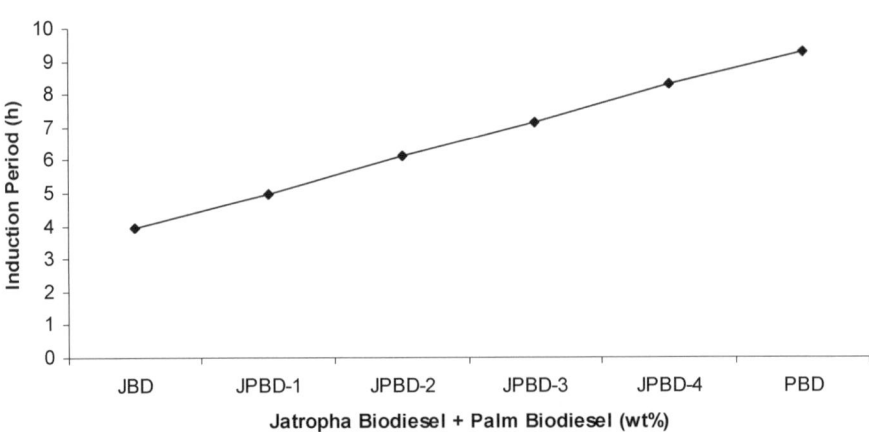

Figure 4.18 The OS of jatropha biodiesel blended with palm biodiesel.

seed oils, and PBD synthesized from edible oil were blended with different weight ratios to examine the influence on the OS (Figures 4.18–4.21).[64] The pongamia, jatropha, and palm biodiesels used in this work had OS values of 2.54, 3.95 and 9.24 h respectively. A minimum dosage of 40 wt% of PBD was needed to improve the IP of neat JBD from 3.95 h to above 6 h as required by the EN 14112 specification for biodiesel OS (Figure 4.18). A minimum dosage of 20 wt% of PBD was needed to improve the IP of neat PoBD from 2.54 h to above 3 h as required by the ASTM D-6751 specification for biodiesel OS, and a minimum dosage of 60 wt% of PBD was needed to improve the IP of neat PoBD from 2.54 h to above 6 h as required by the EN 14112 specification for biodiesel OS (Figure 4.19). A minimum dosage of 40 wt% of JBD was needed to improve the IP of neat PoBD from 2.54 h to above 3 h as required by the ASTM specification for biodiesel OS, but the EN 14214 specification was not

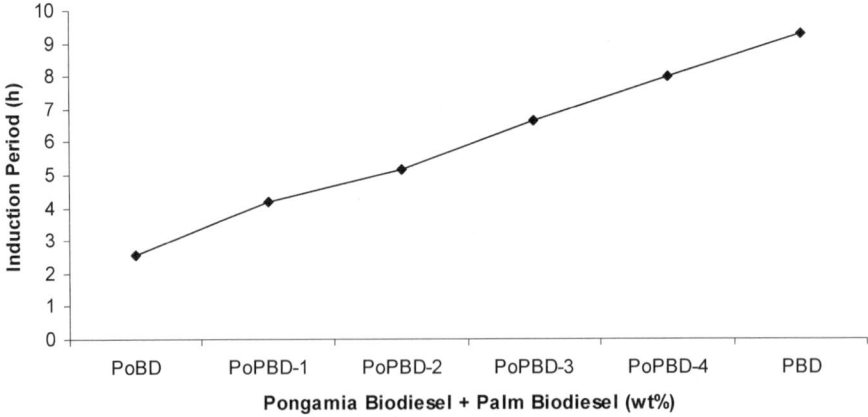

Figure 4.19 The OS of pongamia biodiesel blended with palm biodiesel.

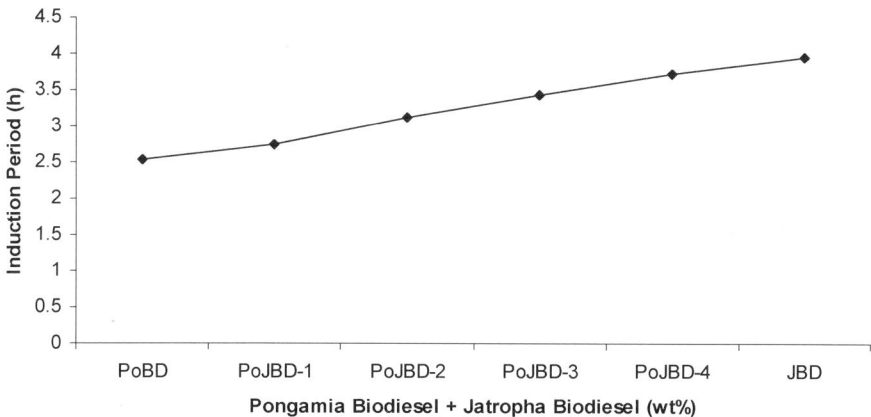

Figure 4.20 The OS of pongamia biodiesel blended with jatropha biodiesel.

fulfilled (Figure 4.20). By blending three biodiesels, low OS of PoBD and JBD was improved (Figure 4.21). ASTM D-6751-08 specification is satisfied for all the biodiesel blends. Therefore, to the OS, blending of two or more than two biodiesels is a simple but effective method.

The palm, rapeseed, and soybean biodiesels used in other work had the following OS values: 11.00, 6.94, and 3.87 h, respectively. When soybean biodiesel was blended with palm and rapeseed biodiesels having higher OS values, the OS of the blended biodiesel was improved.[63]

Researchers also synthesized the surrogate molecules *i.e.*, methyl, ethyl, isopropyl and butyl esters, of β-branched fatty acids which had substantially better OS values, exhibited by IPs of more than 24 h.[87] From these literature reports and quality survey reports,[88–90] it can be concluded that it will not be possible to use biodiesel without antioxidants but the use of antioxidants

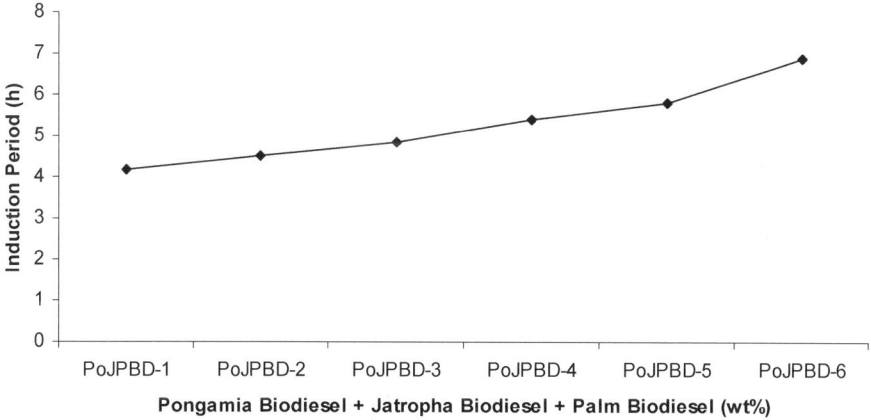

Figure 4.21 The OS of pongamia biodiesel blended with jatropha and palm biodiesels.

only delays the onset of oxidation, *i.e.*, extends the so-called IP until the antioxidant is exhausted and oxidation commences. Thus, fatty materials should not be exposed to oxidation-promoting factors as far as possible even when using antioxidants.

References

1. B. K. Barnwal and M. P. Sharma, *Renewable Sustainable Energy Rev.*, 2005, **9**, 363.
2. A. Murugesan, C. Umarani, R. Subramanian and N. Nedunchezhian, *Renewable Sustainable Energy Rev.*, 2009, **13**, 653.
3. A. A. Refaat, *Int. J. Environ. Sci. Technol.*, 2009, **6**, 677.
4. M. Canakci, A. Monyem and J. Van Gerpen, *Trans. ASAE*, 1999, **42**, 1565.
5. P. Bondioli, A. Gasparoli, L. D. Bella, S. Taghliabue and G. Toso, *Eur. J. Lipid Sci. Technol.*, 2003, **105**, 735.
6. P. Bondioli, A. Gasparoli, A. Lanzani, E. Fedeli, S. Veronese and M. Sala, *J. Am. Oil Chem. Soc.*, 1995, **72**, 699.
7. A. Monyem, M. Canakci and J. Van Gerpen, *Appl. Eng. Agric.*, 2000, **16**, 373.
8. I. Miyata, Y. Takei, K. Tsurutani and M. Okada, SAE Paper No. 2004-01-3031, SAE InternationalWarrendale, 2004.
9. A. Sarin, R. Arora, N. P. Singh, M. Sharma and R. K. Malhotra, *Energy*, 2009, **34**, 1271.
10. A. Sarin, R. Arora, N. P. Singh, R. Sarin, M. Sharma and R. K. Malhotra, *J. Am. Oil Chem. Soc.*, 2010, **87**, 567.
11. A. Sarin, R. Arora, N. P. Singh, R. Sarin and R. K. Malhotra, *Energy Fuels*, 2010, **24**, 2652.
12. S. Jain and M. P. Sharma, *Renewable Sustainable Energy Rev.*, 2010, **14**, 667.
13. R. Sarin, M. Sharma, S. Sinharay and R. K. Malhotra, *Fuel*, 2007, **86**, 1365.
14. G. Knothe, *Fuel Process. Technol.*, 2005, **86**, 1059.
15. R. O. Dunn, *J. Am. Oil Chem. Soc.*, 2002, **79**, 915.
16. G. Knothe, *Energy Fuels*, 2008, **22**, 1358.
17. R. Burton, presented at the Alternative Fuels Consortium, Central Carolina Community College, Sanford, 2008.
18. G. Knothe, *Fuel Process. Technol.*, 2007, **88**, 669.
19. M. W. Formo, E. Jungermann, F. Noris and N. O. V. Sonntag, *Bailey's Industrial Oil and Fat Products*, John Wiley & Sons, New York, 4th edn, 1979, p. 698.
20. J. C. Cowan, Encyclopedia of Chemical Technology, Wiley Interscience, 3rd edn, New York, USA, 1979, p. 130.
21. D. F. Church and W. A. Pryor, *Environ. Health Perspect.*, 1985, **64**, 111.

22. W. E. Neff, T. L. Mounts, W. M. Rinsch and H. Konishi, *J. Am. Oil Chem. Soc.*, 1993, **70**, 163.
23. K. Andersson and H. Lingnert. *J. Am. Oil Chem. Soc.*,1998, **75**, 1041.
24. W. E. Neff, M. A. El-Agaimy and T. L. Mounts, *J. Am. Oil Chem. Soc.*, 1994, **71**, 1111.
25. M. Mittelbach and S. Gangl, *J. Am. Oil Chem. Soc.*, 2001, **78**, 573.
26. P. Bondioli, A. Gasparoli, L. D. Bella and S. Tagliabue, *Eur. J. Lipid Sci. Technol.*, 2002, **104**, 777.
27. L. M. DuPlessis, J. B. M. DeVilliers and W. H. Van der Walt, *J. Am. Oil Chem. Soc.*, 1985, **62**, 748.
28. R. A. Korus, T. L. Mousetis and L. Lloyd, *Am. Soc. Agric. Eng.*, 1982, **18**, 223.
29. E. N. Frankel, *Lipid Oxidation*, The Oily Press, Dundee, 1998, p. 19.
30. Y. H. Hui, in *Bailey's Industrial Oil and Fat Products*, ed. F. Shahidi, John Wiley & Sons, New York, 5th edn, 1996, vol. 4, p. 411.
31. American Oil Chemists' Society Official Method Cd 12b-92, Oil Stability Index, AOCS Press, Champaign, 1999.
32. R. Altin, S. Çetinkaya and H. S. Yücesu, *Energy Convers. Manage.*, 2001, **42**, 529.
33. T. I. Hämäläinen and A. Kamal-Eldin, in *Analysis of Lipid Oxidation*, ed. A. Kamal-Eldin and J. Pokorný, AOCS Press, Champaign, 2005, p. 70.
34. J. Pokorný, S. Schmidt and J. Parkányiová, in *Analysis of Lipid Oxidation*, ed. A. Kamal-Eldin and J. Pokorný, AOCS Press, Champaign, 2005, p. 17.
35. H. L. Fang and R. L. McCormick, SAE Paper No. 2006-01-3300, SAE International, Warrendale, 2006.
36. K. Yamane, K. Kawasaki, K. Sone, T. Hara and T. Prakoso, *Int. J. Engine Res.*, 2007, **8**, 307.
37. R. O. Dunn, *Energy Fuels*, 2008, **22**, 657.
38. L. L. Stavinoha, Report No. SwRI 02-1318-005National Renewable Energy Laboratory, Colorado, USA, 2000.
39. System Lab Services Division, *Determination of Biodiesel Oxidation and Thermal Stability: Final Report*, Williams Pipe Line Company, Tulsa, USA, 1997.
40. I. Hiroaki, M. Eiji H. Shusaku and S. Saka, *Fuel*, 2008, **87**, 1.
41. G. El Diwani and S. El Rafie, *Int. J. Environ. Sci. Technol.*, 2008, **5**, 391.
42. N. T. Joyner and J. E. McIntyre, *Oil Soap*, 1938, **15**, 184.
43. J. C. Cowan, *Ind. Eng. Chem.*, 1949, **41**, 1647.
44. O. C. Johnson and F. A. Kummerow, *J. Am. Oil Chem. Soc.*, 1957, **34**, 407.
45. H. Wexler, *Chem. Rev.*, 1964, **64**, 591.
46. J. Polavka, J. Paligová, J. Cvengroš and P. Šimon, *J. Am. Oil Chem. Soc.*, 2005, **82**, 519.
47. J. F. Reyes and M. A. Sepúlveda, *Fuel*, 2006, **85**, 1714.
48. H. N. Giles, Significance of tests for petroleum products, in ed. S. J. Rand, ASTM International, West Conshohocken, 7th edn, 2003, p. 108.

49. A. Bouaid, M. Martinez and J. Aracil, *Fuel*, 2007, **86**, 2596.
50. D. P. Geller, T. T. Adams, J. W. Goodrum and J. Pendergrass, *Fuel*, 2008, **87**, 92.
51. L. M. Das, D. K. Bora, S. Pradhan, M. K. Naik and S. N. Naik, *Fuel*, 2009, **88**, 2315.
52. E. O. Aluyor, K. O. Obahiagbon and M. Ori-jesu, *Sci. Res. Essays*, 2009, **4**, 543.
53. B. K. Sharma, A. Z. S. Paulo, M. P. Joseph M and Z. E. Sevim, *Fuel Process Technol.*, 2009, **90**, 1265.
54. F. D. Gunstone, *The Chemistry of Fats and Oils*, Blackwell Publishing, Oxford, 2004, p. 150.
55. J. C. Thompson, C. L. Peterson, D. L. Reece and S. M. Beck, *Trans. ASAE*, 1998, **41**, 931.
56. R. L. McCormick, M. Ratcliff, L. Moens and R. Lawrence, *Fuel Process. Technol.*, 2007, **88**, 651.
57. G. Knothe and R. O. Dunn, *J. Am. Oil Chem. Soc.*, 2003, **80**, 1021.
58. P. Bondioli and L. Folegatti, *Riv. Ital. Sostanze Grasse*, 1996, **73**, 349.
59. G. Knothe and K. R. Steidley, *Fuel*, 2005, **84**, 1059.
60. W. Yuan, A. C. Hansen, Q. Zhang and Z. Tan, *J. Am. Oil Chem. Soc.*, 2005, **82**, 195.
61. G. Knothe, *J. Am. Oil Chem. Soc.*, 2002, **79**, 847.
62. W. E. Neff, E. Selke, T. L. Mounts, E. N. Rinsch and M. A. M. Zeitoun, *J. Am. Oil Chem. Soc.*, 1992, **69**, 111.
63. P. Ji-Yeon, K. Deog-Keun, L. Joon-Pyo, P. Soon-Chul, K. Young-Joo and L. Jin-Suk Lee, *Bioresour. Technol.*, 2008, **99**, 1196.
64. A. Sarin, R. Arora, N. P. Singh, R. Sarin and R. K. Malhotra, *Energy*, 2010, **35**, 3449.
65. J. Pospisil and P. P. Klemchuk, *Oxidation Inhibition in Organic Materials*, CRC Press, Boca Raton, 3rd edn, 1990, vol. 1, pp. 1.
66. D. Lomonaco, F. J. N. Maia, C. S. Clemente, J. P. F. Mota, A. E. Costa Junior and S. E. Mazzetto, *Fuel*, 2012, **97**, 552–559.
67. B. R. Moser, *Renewable Energy*, 2012, **40**, 65.
68. R. O. Dunn, *J. Am. Oil Chem. Soc.*, 2005, **82**, 381.
69. R.A. Ferrari, V. da Selva Oliveira and A. Scabio, *Scientia Agricola*, 2005, **62**, 291.
70. M. Mittelbach and S. Schober, *J. Am. Oil Chem. Soc.*, 2003, **80**, 817.
71. H. Tang, H. Wang, S. O. Salley and K. Y. S. Ng, *J. Am. Oil Chem. Soc.*, 2008, **85**, 373.
72. A. Fröhlich and S. Schober, *J. Am. Oil Chem. Soc.*, 2007, **84**, 579.
73. R. O. Dunn, *Fuel Process. Technol.*, 2005, **86**, 1071.
74. Y. C. Liang, C. Y. May, C. S. Foon, M. A. Ngan, C. C. Hock and Y. Basiron, *Fuel*, 2006, **85**, 867.
75. A. K. Domingos, E. B. Saad, W. W. D. Vechiatto, H. M. Wilhelm and L. P. Ramos, *J. Braz. Chem. Soc.*, 2007, **18**, 416.

76. E. Sendzikiene, V. Makareviciene and P. Janulis, *Pol. J. Environ. Stud.*, 2005, **14**, 335.
77. A. Sarin, N. P. Singh, R. Sarin and R. K. Malhotra, *Energy*, 2010, **35**, 4645.
78. S. Jain and M. P. Sharma, *Fuel*, 2011, **90**, 2045.
79. J. L. Reyes-Gavilan and P. Odorisio, A review of the Mechanism of Antioxidants, Metal Deactivators, and Corrosion Inhibitors, Ciba Specialty Chemicals publication, STLE 2002 CD-ROM, May 2002, presented at the National Lubricating Grease Institute Conference, 2002.
80. A. Sarin, R. Arora, N. P. Singh, R. Sarin, R. K. Malhotra, M. Sharma and A. A. Khan, *Energy*, 2010, **35**, 2333.
81. S. G. Morris, J. S. Myers Jr, M. L. Kip and R.W. Riemenschneider, *J. Am. Oil Chem. Soc.*, 1950, **27**, 105.
82. P. R. Krishnamoorthy, S. Vijayakumari and S. Sankaralingam, *IEEE Trans. Electr. Insul.*, 1992, **27**, 271.
83. I. A. Golubeva, E. V. Klinaeva and V. S. Yakovlev, *Chem. Technol. Fuels Oils*, 1994, **30**, 119.
84. I. A. Golubeva, E. V. Klinaeva, V. N. Koshelev, V. I. Kelarev and I. A. Gol'dsher, *Chem. Technol. Fuels Oils*, 1997, **33**, 23.
85. K. Akinori, *Petrotech*, 2007, **30**, 63.
86. R. Sarin, R. Kumar, B. Srivastav, S. K. Puri, D. K. Tuli, R. K. Malhotra and A. Kumar, *Bioresour. Technol.*, 2009, **100**, 3022.
87. R. L. McCormick, T. L. Alleman, M. Ratcliff and L. Moens, Technical Report No. NREL/TP-38836National Renewable Energy Laboratory, Colorado, USA, 2004.
88. T. L. Alleman and R. L. McCormick, National Renewable Energy Laboratory, NREL/TP-540-42787, Colorado, USA, 2008.
89. H. Tang, N. Abunasser, A. Wang, B. R. Clark, K. Wadumesthrige, S. Zeng, H. Kim, S. O. Salley, G. Hirschlieb, J. Wilson and K.Y. S. Ng, *Fuel*, 2008, **87**, 2951.

CHAPTER 5

Low-Temperature Flow Properties of Biodiesel

5.1 Introduction

Although biodiesel is environmentally compatible, it has certain limitations at low temperatures. The low-temperature flow properties of biodiesel are characterized by its cloud point (CP), pour point (PP) and cold filter plugging point (CFPP) and these must be considered when operating compression–ignition engines in moderate-temperature climates during winter months. The CP is the temperature at which a sample of the fuel starts to appear cloudy, indicating that wax crystals have begun to form which can clog the fuel lines and filters in a vehicle's fuel system, the PP is the temperature below which the fuel will not flow, and the CFPP is the temperature at which fuel causes a filter to plug as a result of its crystallization.[1-4] Neither CP nor PP efficiently predicts how diesel fuels perform in tanks and fuel systems during cold weather, therefore, low-temperature filterability tests like CFPP were developed that more effectively predict limiting fuel temperatures where start-up or operability problems may be expected after prolonged exposure. The CFPP test is accepted nearly worldwide. There is another test method for the low-temperature flow properties of conventional diesel fuel, namely the low-temperature flow test (LTFT) used in North America (ASTM D-4539).[5] This method has also been used to evaluate biodiesel and its blends with Nos. 1 and 2 conventional diesel fuel. In recent years, another low-temperature operability problem has been recognized, resulting from the formation of insoluble particles upon storage at cool temperatures—though generally above the CP.[3] These insolubles arise from the precipitation of trace-level non-fatty acid methyl ester (FAME) impurities, not from the major FAME components

Biodiesel: Production and Properties
Amit Sarin
© Amit Sarin 2012
Published by the Royal Society of Chemistry, www.rsc.org

themselves. The two major families of impurities identified as causing such precipitation problems are saturated monoglycerides and sterol glucosides.[6–11] The test developed to assess this problem is known as the cold soak filterability test (CSFT). The European EN and USA ASTM standards include the low-temperature flow properties (discussed in Chapter 3) but limits are not given, rather a report is required. This is due to the varying weather conditions in different countries. As CP is the trigger for negative effects on fuel injection, its prediction is extremely meaningful. While most of the properties of biodiesel are comparable to petroleum-based diesel fuel, improvement of its low-temperature flow properties still remains one of the major challenges while using biodiesel as an alternative fuel for diesel engines.

5.2 Factors Affecting the Low-Temperature Flow Properties of Biodiesel

5.2.1 Biodiesel Composition

As already mentioned in Chapter 4, the properties of vegetable oils depend upon the fatty acid composition of the oil. Common fatty acids with their chemical structures are shown in Table 4.1. Several researchers have studied the CP and PP of neat biodiesel and biodiesel blends. It was found that the CP of biodiesel could be determined solely from the amount of saturated FAMEs, regardless of the composition of the unsaturated esters.[3] On the other hand, soybean oil methyl ester is composed of 15–18 wt% total methyl palmitate and stearate contents and has a CP = 0, PP = −2, and CFPP = −2 °C.[1] Poor low-temperature flow properties of biodiesel can be explained by a high content of saturated fatty acid alkyl esters because the unsaturated fatty acid alkyl esters have lower melting points than the saturated fatty acid alkyl esters.[5] Poor low-temperature flow properties result from the presence of long-chain, saturated FAMEs in biodiesel. Saturated methyl esters longer than C_{12} significantly increase the CP and PP, even when blended with conventional diesel.[12,13] In general, the longer the carbon chain, the higher the melting point, and the poorer the low-temperature properties.[1,14–16]

The key feature of the fatty acid moiety is the length of the fatty chain. A strong correlation is observed between the chain length and the melting point of the corresponding fatty acid (FA).[5] Table 5.1 summarizes literature values for the melting points of some saturated fatty acids and their esters.[5]

There is a clear trend of increasing melting point with increasing chain length, evident for both the fatty acid and the ester. The complex nature of natural fats and oils ensures that any elucidation of such trends from the acid profiles is difficult. Coconut oil contains very high levels of saturated, low-molecular-weight fatty acids and relatively few unsaturated FAs, therefore its ethyl ester possess a relatively high CP of 5 °C.[17] Safflower oil ethyl ester has a very high proportion of unsaturated FAs, with few saturated FAs and its CP is −6 °C.[17] Safflower and sunflower oil have similar proportions of saturation

Table 5.1   Melting points of acids and esters.

Trivial (systematic) name; ratio	Melting point/°C
Caprylic (octanoic) acid; 8 : 0	16.5
Ethyl ester	−43.1
Capric (decanoic) acid; 10 : 0	31.5
Ethyl ester	−20
Lauric (dodecanoic) acid; 12 : 0	44
Methyl ester	5
Ethyl ester	−1.8
Myristic (tetradecanoic) acid; 14 : 0	58
Methyl ester	18.5
Ethyl ester	12.3
Palmitic (hexadecanoic) acid; 16 : 0	63
Methyl ester	30.5
Ethyl ester	19.3/24
Stearic (octadecanoic) acid; 18 : 0	71
Methyl ester	39
Ethyl ester	31–33.4

but sunflower ethyl ester has a CP of −1 °C.[17] The nature of the saturated and unsaturated fractions results in a much higher CP for sunflower ethyl ester. Firstly, the fraction of C18 : 0 and C20 : 0 in sunflower oil is almost double that for safflower oil. Thus, the longer chain length of the saturated fraction results in a higher CP for sunflower ethyl ester. Secondly, the unsaturated portion of sunflower oil contains less of the highly non-linear C18 : 2 and more C18 : 1 than safflower oil. This large diversion in CP demonstrates the complex inter-relationship between chain length and saturation level. It is therefore evident that very small differences in the proportions of fatty acids can have a large impact on cold-flow properties.

5.2.2   Alcoholic Head-Group

The nature of the alcoholic head-group of the ester can also influence the low-temperature flow properties of biodiesel. The use of long (3–8 carbon) normal alcohols or branched alcohols to manufacture biodiesel reduces the CP compared to those for conventional methyl esters. Methyl fatty acids possess polarity in their head-groups to provide an amphiphilic nature that results in the head-to-head alignment of molecules.[18] Ethyl or larger alkyl esters include non-polar head-groups that are sufficiently large to shield the forces between more polar portions of the head-group. These esters therefore align themselves in a head-to-tail arrangement with much larger molecular spacing. Bulky head-groups disrupt the spacing between individual molecules in the crystal structure causing rotational disorder in the hydrocarbon tail-group.[19] The effective result is that melting points for alkyl palmitate and stearate esters exhibit a decline for head-group lengths up to and including *n*-butyl, but then increase for *n*-pentyl and larger.

Esters with branched head-groups show reductions in CP of around 3 °C for canola oil and up to 9 °C for soybean oil.[19] It was discovered that isopropyl and 2-butyl esters of normal soybean oil crystallized 7 to 11 and 12 to 14 K lower, respectively, than the corresponding methyl esters.[7] Isopropyl esters of lard and tallow had crystallization temperatures similar to those for the methyl esters of soybean oil despite their increased saturation level.

5.2.3 Influence of Minor Components

As discussed in Chapter 2, biodiesel is generally produced by transesterification. The process generally follows three steps where each fatty acid group is sequentially removed and converted into a fatty acid alkyl ester (biodiesel) molecule. Thus, at complete conversion, one mole of triglycerides will generate three moles of biodiesel. This process may leave behind very small concentrations of minor constituents such as residual oil, glycerides or fatty acids.[20] Some constituents may be present in the crude feedstock and be carried through after conversion to biodiesel. Other constituents may be present due to incomplete conversion and processing of the biodiesel product. These materials have high melting points and very low solubilities allowing them to form solid residues when stored during cold weather. Settling solid residues were found to clog fuel filters in fuel dispensers and vehicles. Thus, these materials can influence the low-temperature flow properties, OS and other physicochemical properties of biodiesel.

Another class of components found to naturally occur in plant oils is steryl glucosides. These compounds are glycosides derived from a phytosterol and glucose (Figure 5.1).

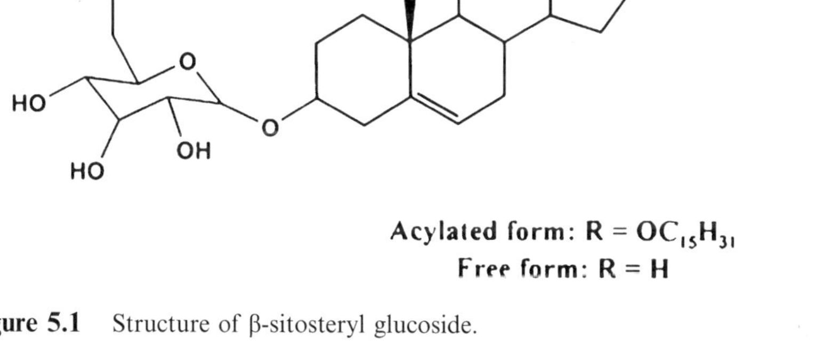

β-Sitosteryl Glucoside

Acylated form: R = $OC_{15}H_{31}$
Free form: R = H

Figure 5.1 Structure of β-sitosteryl glucoside.

They exist mostly in acylated form, which is very soluble in vegetable oils. Depending on processing, total steryl glucoside content may be higher in soybean oil than other plant oils.[21] Processing before and after conversion to biodiesel may reduce the steryl glucoside content such that they pose no significant problems. However, if residual steryl glucosides are present at low concentrations, they can settle at the bottom of storage tanks.[21] The presence of moisture can enhance problems because both monoglycerides (MGs) and diglycerides (DGs) are amphiphilic in nature and may interact with water. Low-temperature conditions can also enhance the problem and may cause particles to act as seed crystals for forming larger agglomerates.[20,21]

Liquid and solid samples collected from palm oil FAMEs and soybean methyl ester storage tanks were evaluated by nuclear magnetic resonance (NMR) and gas chromatography-mass spectroscopy.[22] Samples were distilled to reduce the FAME content and concentrate the heavier constituents such as MGs. Analysis of hazy FAME samples indicated trace concentrations of acylated steryl glucosides plus, MGs and DGs.

The contents of solid residues from commercial soybean methyl ester, cotton seed oil FAMEs and poultry fat FAMEs and their blends with ultra-low sulfur petrodiesel were analyzed by gas chromatography (GC) and Fourier transfer infrared (FT-IR) spectroscopy.[9] Samples were blended at B20 and chilled at 4 °C for 24 h to promote precipitate formation. After isolating the solids, FT-IR and GC analyses revealed mainly steryl glucosides in precipitates from soybean methyl ester (SME). In contrast, precipitates from poultry fat FAMEs were mainly composed of saturated MGs. Cotton seed oil FAME precipitates were composed of steryl glucosides and saturated MGs.

There are a number of reports of the influence of the aforementioned on the low-temperature flow properties of biodiesel.[23–25] As early as 1996, MGs and DGs were detected in a creamy paste material recovered from plugged fuel filters in a bus refueling station in Ames, Iowa.[23] The buses were running on a B20 biodiesel blend in petrodiesel. Analysis of the paste revealed 25 wt% MGs + DGs, where the MGs were 95% saturated.

The effects of trace concentrations of MGs, steryl glucosides, water and soap and combinations of these constituents on the CP of soybean methyl ester was determined.[24] Results from a series of statistically designed experiments showed that the maximum increase in CP was 3.6 °C for soybean methyl ester doped with 1 wt% Monoacylglycerol (MAG), 40 ppm steryl glucosides, 40 ppm soap and 500 ppm water. Individually, the most significant effects on increasing CP were caused by adding 0.6% MGs. The presence of steryl glucosides did not greatly affect the CP.

Studies have compared the cold-flow properties of biodiesel/diesel blends, and found a tendency for these blends to form solid residues while being stored at 4 °C for 24 h.[7,9] Results for blends with soybean methyl ester, palm oil FAMEs, cotton seed oil FAMEs, poultry fat FAMEs and yellow grease FAMEs showed that the CFPP was very close to the CP at low blend ratios and tended to converge close to the PP at higher blend ratios. Blends with a higher tendency to form solid residues during cold storage demonstrated CFPP–PP convergence at blend ratios between B10 and B20. This was the case for yellow grease FAMEs, cotton seed oil FAMEs and one of two soybean methyl ester (SME)-based blends studied. In contrast, palm oil FAMEs, poultry fat FAMEs and the second soybean methyl ester-based blends showed a lower tendency to form solid crystals after cold storage. This observation explained why these blends showed CFPP–PP convergence at higher (B50–B70) blend ratios.

The numerous cases of clogged diesel fuel filters stressed a need to develop test methods for analyzing the potential of biodiesel to form solid precipitates after storage in cold temperatures.

In response, the biodiesel industry collaborated with ASTM to improvise a specialized test to identify fuels that have the potential to clog filters.[14] CSFT limits and methods are discussed in Chapter 3. Nevertheless, the CSFT is viewed by many biodiesel producers as flawed for several reasons.[26] First, maximum time limits as specified in D-6751 were empirically set based on anecdotal evidence, and have little scientific basis. The test is not a real analytical test and demonstrates poor reproducibility and repeatability. It takes up to 36 h to complete, a period that is too long for many producers.

Regardless of origin, large quantities of solid particles settling at the bottom of storage tanks can restrict flow or clog filters in fuel dispensers and vehicles after transfer at temperatures above the CP of the fuel. Thus, maintaining fuel quality year-round is becoming more of an imperative for the biodiesel industry as commercial markets continue to grow worldwide.

5.3 Statistical Relationship between Low-temperature Flow Properties and Fatty Acid Methyl Ester Composition

As discussed above, the properties of the various individual fatty esters that comprise biodiesel determine its overall fuel properties, and in turn, the properties of various fatty esters are determined by their structural features. Researchers have also developed a statistical relationship between low-temperature flow properties and the FAME content.[27,28]

Palm biodiesel (PBD), jatropha biodiesel (JBD) and pongamia biodiesel (PoBD) were blended with different weight ratios (%) as follows: 100 : 00 : 00, 80 : 20 : 00, 80 : 00 : 20, 60 : 40 : 00, 60 : 20 : 20, 60 : 00 : 40, 40 : 60 : 00, 40 : 40 : 20, 40 : 20 : 40, 40 : 00 : 60, 20 : 80 : 00, 20 : 60 : 20, 20 : 40 : 40, 20 : 20 : 60, 20 : 00 : 80, 00 : 100 : 00, 00 : 80 : 20, 00 : 60 : 40, 00 : 40 : 60, 00 : 20 : 80, 00 : 00 : 100 respectively.[28] Their fatty acid profiles were determined, as shown in Table 4.3. The aforementioned 21 biodiesel samples were used for the experiment. The FAME compositions of PBD, JBD and PoBD samples (Table 5.2) and their 21 blends were determined by GC using nitrogen as the carrier gas and di(ethylene glycol) succinate.

Table 5.2   Fatty acid methyl ester compositions of biodiesel samples.

Fatty acid methyl ester	JBD/wt%	PoBD/wt%	PBD/wt%
Palmitic (C16 : 0)	14.2	9.8	40.3
Palmitolic (C16 : 1)	1.4	—	—
Stearic (C18 : 0)	6.9	6.2	4.1
Oleic (C18 : 1)	43.1	72.2	43.4
Linoleic (C18 : 2)	34.4	11.8	12.2
Saturated	21.1	16.0	44.4
Unsaturated	78.9	84.0	55.6

Palm, jatropha, and pongamia biodiesels had the following CPs: 16.0, 4.0, and -1.0 °C respectively and PPs of 12.0, -3.0, and -6.0 °C respectively.[28] Therefore, the order of CP and PP was PBD $>$ JBD $>$ PoBD. Palm, jatropha, and pongamia biodiesels had CFPPs of 14, 1, and -2 °C respectively.[27] Therefore, the order of CFPP was PBD $>$ JBD $>$ PoBD.

The FAME compositions of the three biodiesel samples given in Table 5.2 clearly indicate the predominance of saturated FAMEs in PBD. JBD mainly consisted of esters of oleic and linoleic acids. PoBD contained mainly esters of oleic acid. Data for the CP, PP and CFPP and total FAME composition for all biodiesel samples were imported into SPSS for Windows, version 16.0 software for statistical regression analysis.

Figure 5.2 shows that the content of palmitic acid methyl ester (PAME) has a high degree of correlation with the CP of the three biodiesel blends.[28] Correlation between the CP and PAME content (wt%) was obtained as follows:

$$CP = 0.526(PAME) - 4.992 \ (0 < PAME < 45) \tag{1}$$

For this equation, the coefficient of correlation (R) is 0.981, the coefficient of determination (R^2) is 0.963 and the standard error of estimate (σ_{est}) is 0.929.

A good correlation was also obtained from Figure 5.3 between the PP and PAME content (wt%) of three biodiesel blends as follows:

$$PP = 0.571(PAME) - 12.240 \ (0 < PAME < 45) \tag{2}$$

For this equation, $R = 0.929$, $R^2 = 0.863$ and $\sigma_{est} = 2.039$.

Since PAME was the dominant component of the saturated FAMEs in the biodiesel blends, a similar correlation of CP and PP could be found for the saturated FAMEs (figures not shown).

A good correlation could not be found for variation of the CP and PP with the content of oleic acid methyl ester (OAME) of biodiesel blends, respectively (Figures 5.4 and 5.5).[28] Similarly, a good correlation could not be found for variation of the CP and PP with the content of linoleic acid methyl ester (LAME), respectively (Figures 5.6 and 5.7). This result showed that with

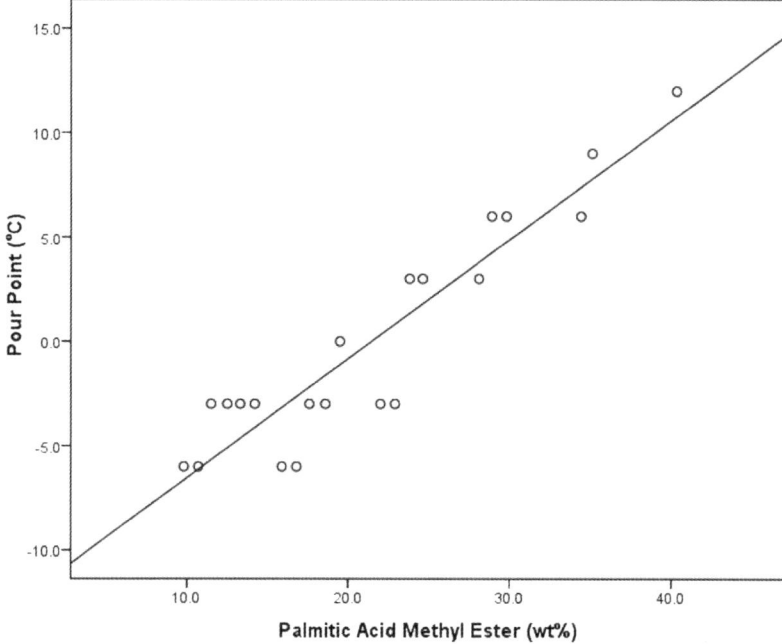

Figure 5.2   The CP of blended biodiesels with the content of palmitic acid methyl ester.

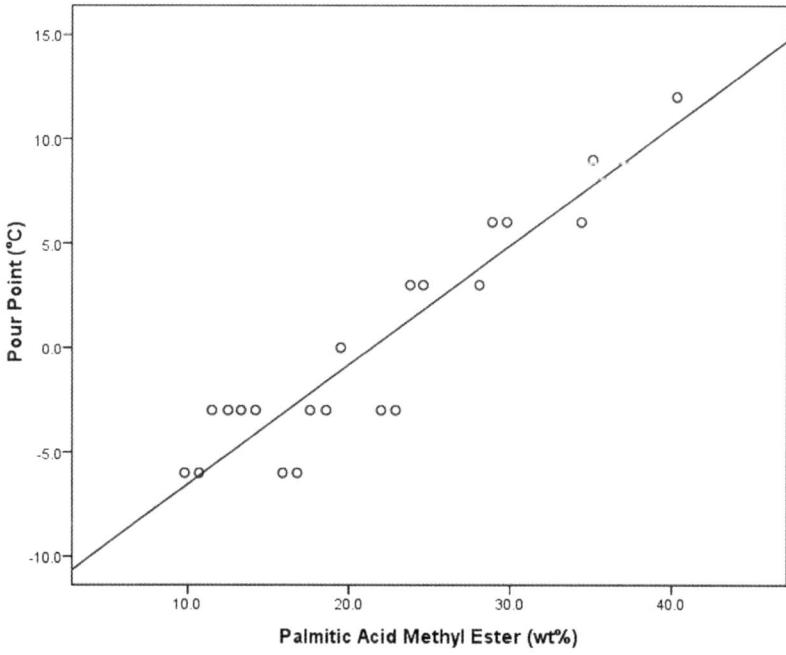

Figure 5.3   The PP of blended biodiesel with the content of palmitic acid methyl ester (PAME).

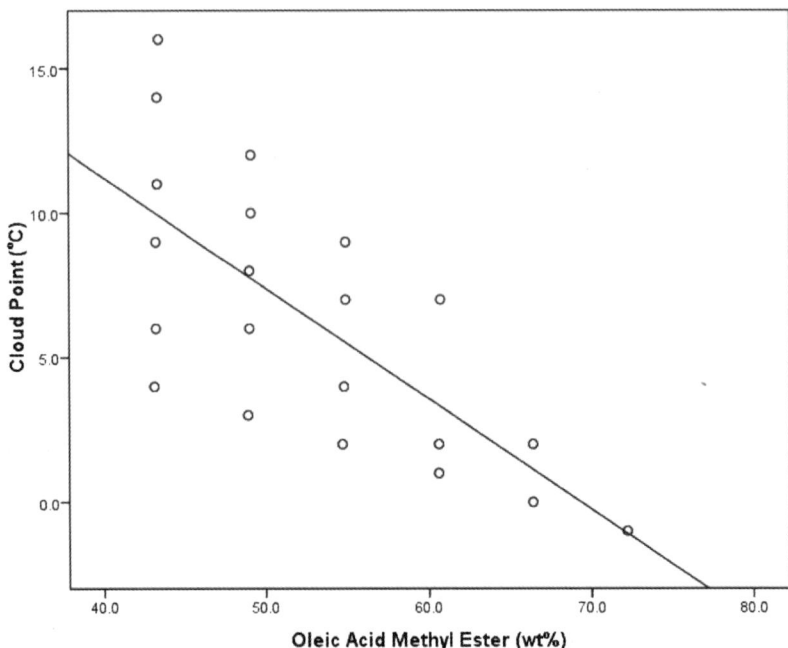

Figure 5.4   The CP of blended biodiesels with the content of oleic acid methyl ester.
$R = 0.722$, $R^2 = 0.521$ and $\sigma_{est} = 3.32$.

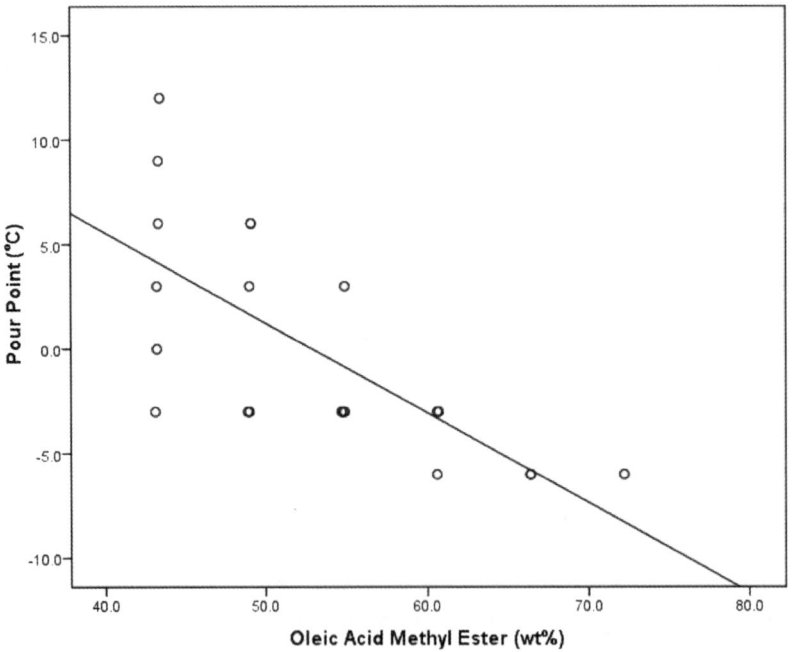

Figure 5.5   The PP of blended biodiesels with the content of oleic acid methyl ester.
$R = 0.709$, $R^2 = 0.502$ and $\sigma_{est} = 3.88$.

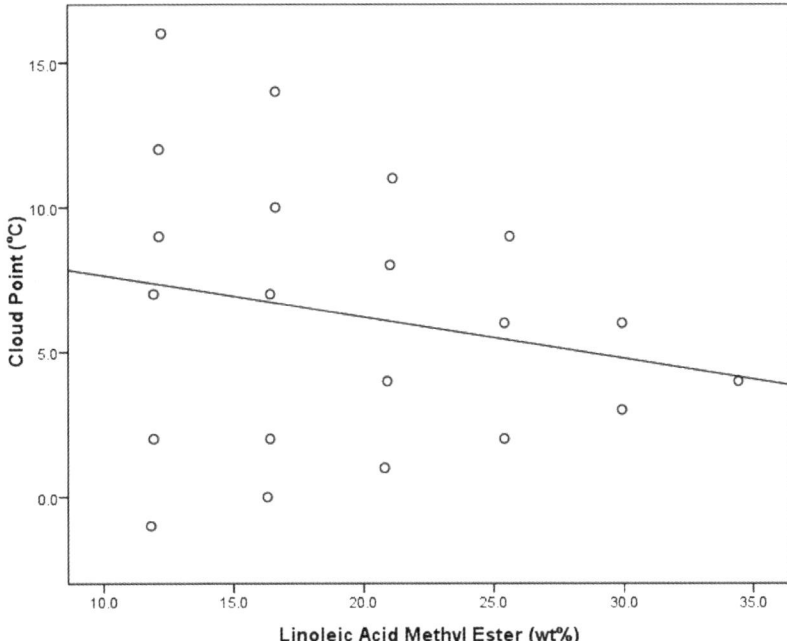

Figure 5.6 The CP of blended biodiesels with the content of linoleic acid methyl ester. $R = 0.208$, $R^2 = 0.043$ and $\sigma_{\text{est}} = 4.69$.

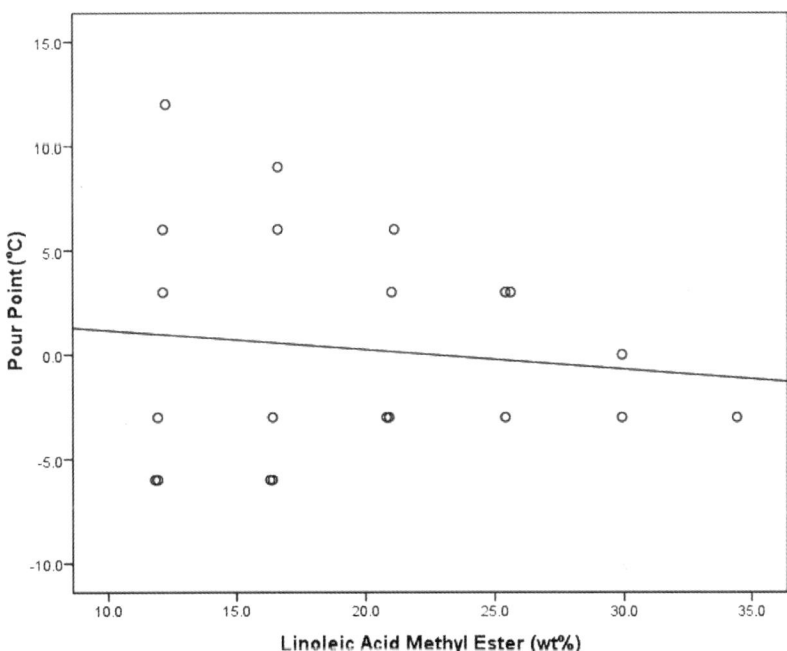

Figure 5.7 The PP of blended biodiesels with the content of linoleic acid methyl ester. $R = 0.119$, $R^2 = 0.014$ and $\sigma_{\text{est}} = 5.46$.

regard to the effects of the individual unsaturated FAMEs, good correlations could not be found.

However, when the effects of the total unsaturated FAME content on the CP were determined, a high degree of correlation was found (Figure 5.8).[28] The correlation is given as follows:

$$CP = 0.576(X) + 48.255 \ (0 < X \leq 84) \tag{3}$$

where X is the content of total unsaturated FAMEs (wt%). For this equation, $R = 0.986$, $R^2 = 0.973$ and $\sigma_{est} = 0.788$.

A good correlation was also obtained from Figure 5.9 between the PP and the content of total unsaturated FAMEs of the three biodiesel blends as follows:

$$PP = 0.626(X) + 45.594 \ (0 < X \leq 84) \tag{4}$$

For this equation, $R = 0.935$, $R^2 = 0.874$ and $\sigma_{est} = 1.955$.

Therefore when the compositions of blended biodiesels are known, the CP and PP can be predicted by using eqns (1)–(4). For example, when the weight ratio of palm, jatropha, and pongamia biodiesels is 60 : 20 : 20, the content of PAME is 28.9 wt%, the CP calculated from eqn (1) is 10.21 °C which is very

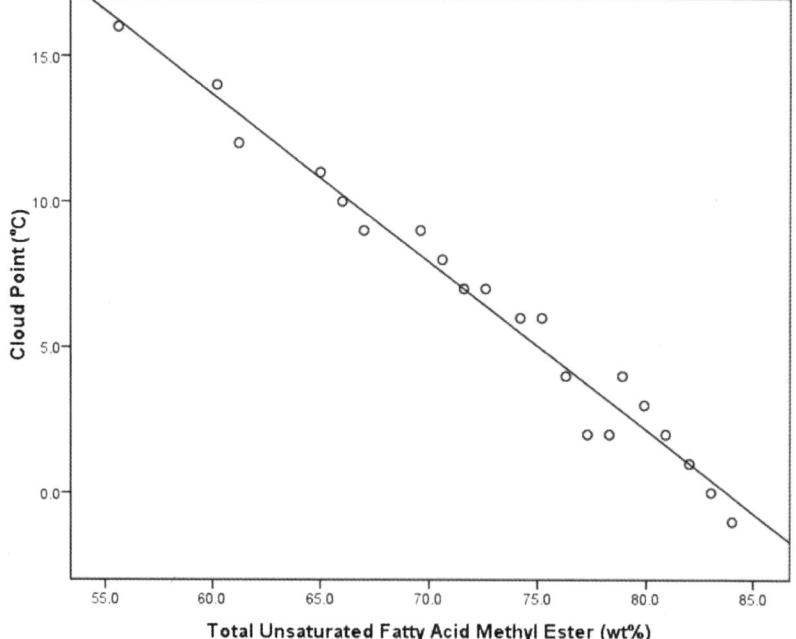

Figure 5.8   The CP of blended biodiesels with the content of total unsaturated fatty acid methyl ester.

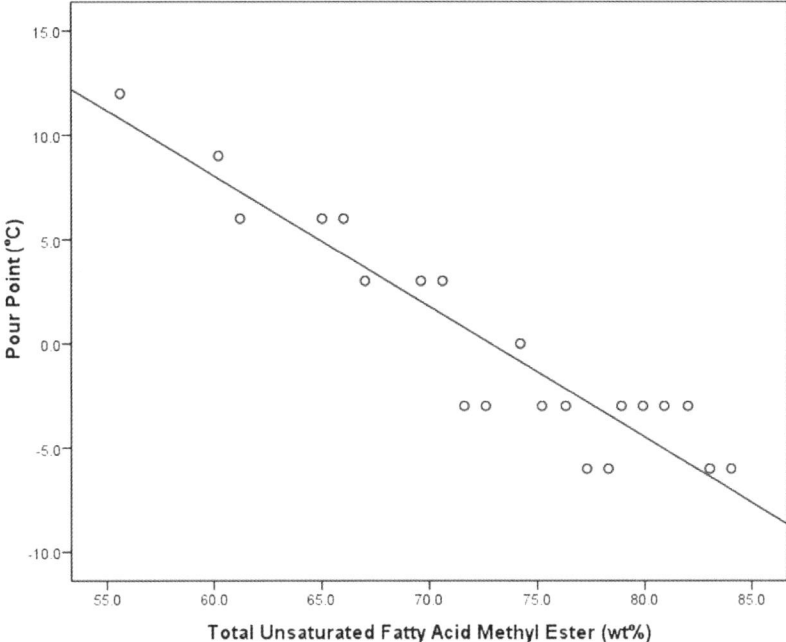

Figure 5.9 The PP of blended biodiesels with the content of total unsaturated fatty acid methyl ester.

close to the experimental value of 10 °C (Figure 5.2). The minimum error of prediction in CPs calculated from eqn (1) is 0.05 and the maximum error is 1.84. The PP determined from eqn (2) is 4.46 °C which is also close to the experimentally determined value of 6 °C (Figure 5.3). The minimum and maximum error of prediction in PPs calculated from eqn (2) is 0.13 and 3.83 respectively. Similarly, the CP determined from eqn (3) is 10.41 °C and the PP from eqn (4) is 4.46 °C for biodiesel blends of weight ratios 60 : 20 : 20, when the content of total unsaturated FAMEs is 65.7 wt%. The minimum and maximum error of prediction in CPs calculated from eqn (3) is 0.013 and 1.73 respectively. The error of prediction in PPs calculated from eqn (4) is a minimum of 0.36 and maximum of 3.77 respectively. Therefore, using eqns (1)–(4), it is possible to directly predict the CPs and PPs of biodiesel blends from the content of PAME as well as from the total unsaturated FAME content.

The dependence of the CFPP on FAME content was also determined.[27] A good correlation was obtained between the CFPP and PAME content (wt%) of three biodiesel blends as follows (Figure 5.10):

$$CFPP = 0.511(PAME) - 7.823 \ (0 < PAME < 45) \tag{5}$$

For this equation, $R = 0.929$, $R^2 = 0.863$ and $\sigma_{est} = 1.831$.

Since PAMEs were the dominant component of saturated FAMEs in the biodiesel blends, a similar correlation of CFPP could be found in the saturated FAMEs (figure not shown).

A good correlation could not be found for variation of the CFPP with contents of OAMEs of biodiesel blends, respectively (Figure 5.11). Similarly, a good correlation could not be found for variation of the CFPP with the contents of LAMEs, respectively (Figure 5.12).

This result showed that with regard to the effects of the individual unsaturated FAMEs, good correlations could not be found. However, when the effect of contents of total unsaturated fatty acid methyl ester (X) composition on the CFPP was determined, a good negative correlation was obtained between the CFPP and X of three biodiesel blends as follows (Figure 5.13):

$$CFPP = -0.561(X) + 43.967 \; (0 < X \le 84) \qquad (6)$$

For this equation, $R = 0.934$, $R^2 = 0.872$ and $\sigma_{est} = 1.762$.

Therefore when the compositions of blended biodiesels are known, the CFPP can be predicted by using eqns (5) and (6). For example, when the weight ratio of palm, jatropha, and pongamia biodiesels is 40 : 40 : 20, the content of PAME is 23.8 wt%, and the CFPP determined from eqn (5) is

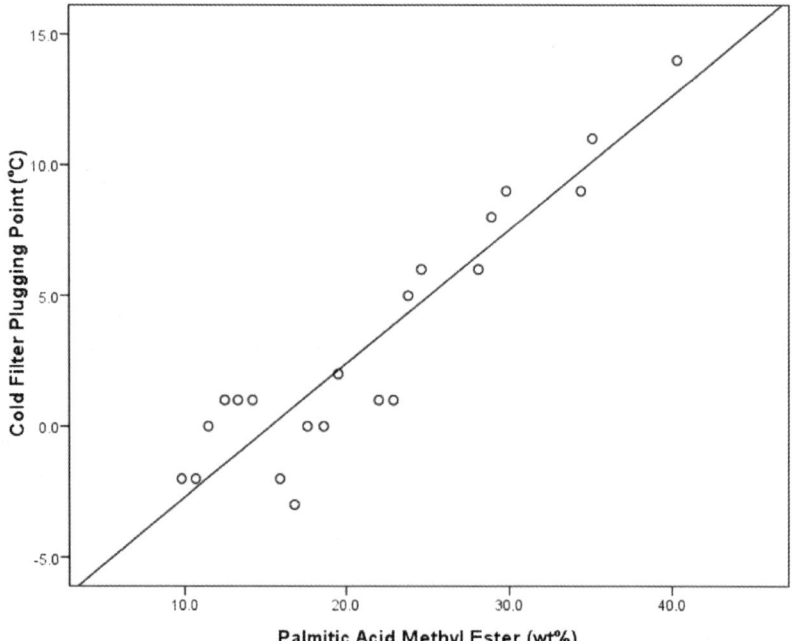

Figure 5.10   The CFPP of blended biodiesels with the contents of palmitic acid methyl ester.

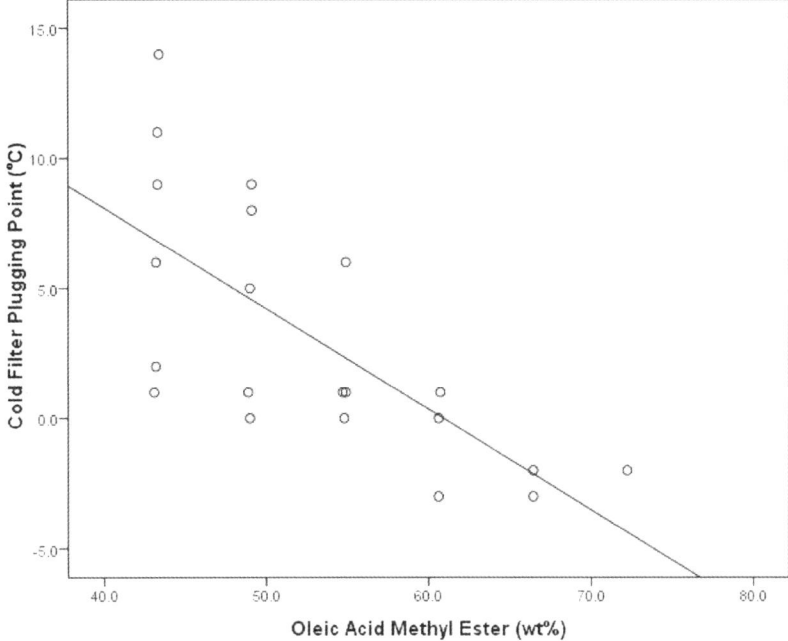

Figure 5.11 The CFPP of blended biodiesels with the contents of oleic acid methyl ester. $R = 0.703$, $R^2 = 0.495$ and $\sigma_{\text{est}} = 3.5494$.

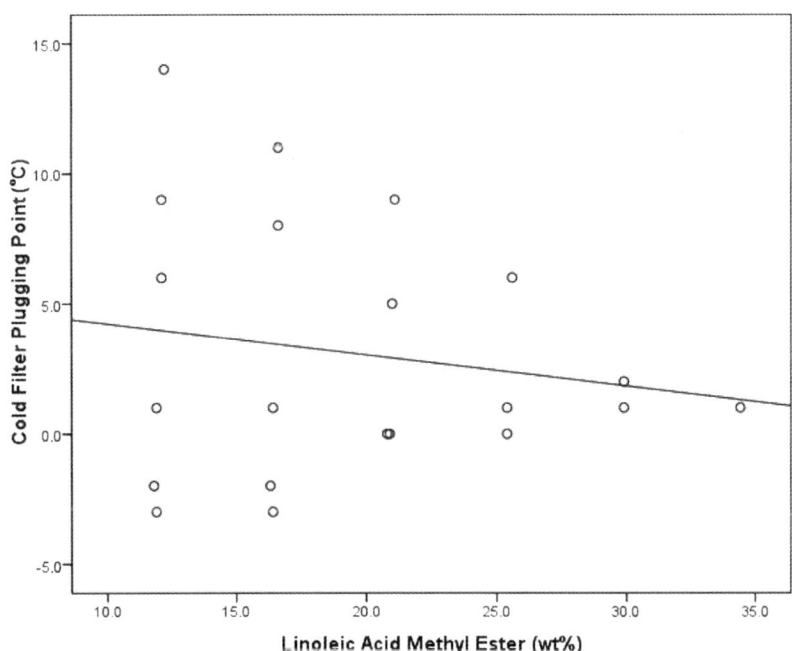

Figure 5.12 The CFPP of blended biodiesels with the contents of linoleic acid methyl ester. $R = 0.168$, $R^2 = 0.028$ and $\sigma_{\text{est}} = 4.9226$.

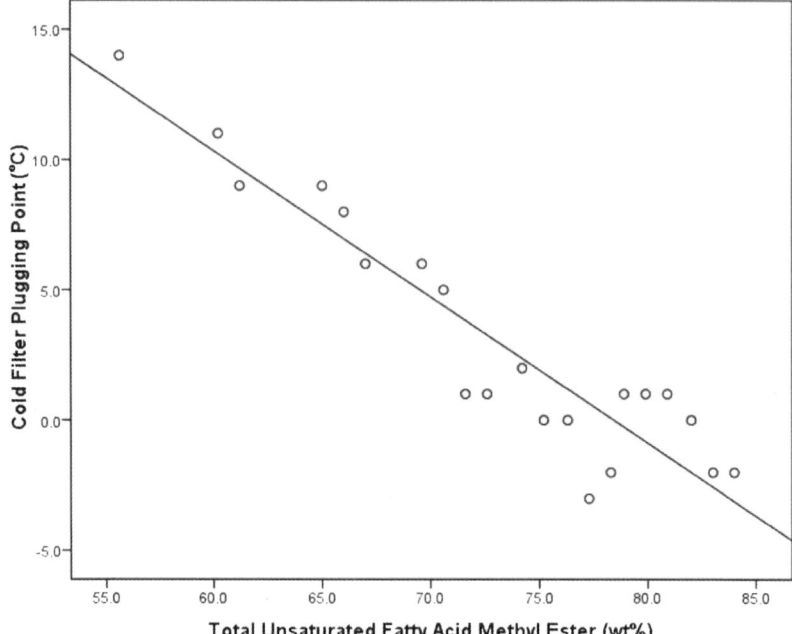

Figure 5.13   The CFPP of blended biodiesels with the contents of total unsaturated fatty acid methyl ester.

4.34 °C which is also close to the experimentally determined value of 5 °C (Figure 5.10). The minimum and maximum error of prediction in CFPP calculated from eqn (5) is 0.14 and 3.76 respectively.

Similarly, the CFPP from eqn (6) is 4.36 °C for biodiesel blends of weight ratios 40 : 40 : 20, when X is 70.6 wt%. The error of prediction in CFPPs calculated from eqn (6) is a minimum of 0.34 and a maximum of 3.6. Therefore, using eqns (5) and (6), it is possible to directly predict the CFPP of biodiesel blends from the contents of PAMEs as well as from the total unsaturated FAMEs.

In another study to correlate fatty acid profiles with cold-flow properties, by other researchers, blends of palm, rapeseed and soybean oil methyl esters created 21 different FAMEs.[29] The CFPPs of the resulting blends were determined. The researchers were unable to discover a direct correlation between the CFPP and the oleic acid (C18 : 1) content, but they found a strong positive relationship with palmitic acid (C16 : 0) content and a strong negative relationship with the total fraction of unsaturated fatty acids. The correlation between the CFPP and the total fatty acid contents was obtained as:[29]

$$Y = -0.4880X + 36.0548 \ (0 < X \leq 88) \text{ and}$$

$$Y = -2.7043X + 232.0036 \ (88 < X < 100)$$

where X is the content of total unsaturated fatty acids (wt%) and Y is the CFPP.

A thermodynamic study of binary mixtures of various FAMEs was performed to develop a prediction model for estimating the CP of biodiesel produced from various feedstocks.[8] The methyl esters of linseed, safflower, sunflower, rapeseed, soybean, olive, palm and beef tallow were used. It was concluded that the CP of biodiesel could be determined mainly from the amount of saturated fatty acid methyl esters, regardless of the chemical nature of unsaturated esters. Binary mixtures of a saturated and an unsaturated FAME such as methyl palmitate (C16 : 0) and methyl oleate (C18 : 1) produced a monotonic increase in CP with increasing fraction of C16 : 0, from the CP of pure C18 : 1 to that of pure C16 : 0. For mixtures containing only saturated esters (C16 : 0 and C18 : 0 for instance), a lower CP at the eutectic point was observed for a particular composition below the CP of either pure component. The comparison of the CP of the binary mixtures with the so-called CP of the pure components is however not a valid one since the CP of pure compounds is actually better described as the melting point.[18] In another scenario, multi-component mixtures of four different esters were prepared and the CP of those was determined to elucidate trends.[18] When the amount of saturated esters was fixed and only the fractions of unsaturated esters was changed, that is, the ratio of mono- and di-unsaturated ester (C18 : 1 and C18 : 2), the CP remained practically unchanged. By contrast, if the total amount of saturated esters changed, then a dramatic change in CP was observed. Finally, the thermodynamic model was compared with the CP of FAMEs produced from real feedstocks. The discrepancy between the predicted CP for the eight different fats/oils and the actual CP ranged from 0 to 5 K. Clearly, further work is required to improve the accuracy of the model. More researchers have developed predictive models for the CP, based upon these thermodynamic relationships.[30,31] In general, these models show good agreement with laboratory measurements.

With the help of correlations between biodiesel low-temperature flow properties and FAME composition, the low-temperature flow properties of biodiesels produced from new resources can be easily estimated.

There are also empirical correlations of low-temperature flow properties with other factors. Empirical correlations were developed to calculate the CP with respect to the total glycerol content in blends of soybean methyl ester, canola oil FAMEs and ethyl esters from soybean and canola oils in petrodiesel.[32] Although coefficients of regression were good ($R^2 = 0.93$) for soybean methyl ester (SME) and soybean oil fatty acid ethyl esters, they were not as strong (0.78 and 0.53) for blends with canola oil fatty acid alkyl esters. Researchers also developed the distinct artificial neural networks model for the prediction of the CP and PP.[33] Empirical relations to predict both the CP and PP of biodiesel and its blends with No. 2 diesel fuel have been developed as:[34]

$$T_{CP} = 256.4 + 0.1991 V_B + 0.000223 V_B^2$$

$$T_{PP} = 253.9 + 0.1865 V_B + 0.000335 V_B^2$$

where T_{CP} and T_{PP} are the CP and PP respectively and V_B is the volume fraction (%) of biodiesel in the blend.

5.4 How to Improve the Low-temperature Flow Properties of Biodiesel

Several methods have been investigated to improve the low-temperature flow properties of biodiesel, such as by blending with petroleum diesel, the blending of different biodiesels, the use of additives, and the chemical or physical modification of either the oil feedstock or the biodiesel product.

5.4.1 Additives

It was found that blending of biodiesel with petroleum diesel is only effective at low biodiesel proportions (up to 30 vol%) with CPs up to *ca.* −10 °C.[2,35] It was concluded that the CP and PP decrease with an increase in concentration of petroleum diesel in the blend.[34] The empirical relation for this was mentioned earlier in this chapter. Researchers determined the low-temperature properties of monoalkyl esters derived from tallow and recycled greases for neat esters and 20% ester blends in No. 2 low-sulfur diesel fuel.[15] The secondary alkyl esters of tallow showed significantly improved cold-temperature properties over the normal tallow alkyl ester derivatives.

The use of additives is further divided into two types. The first are traditional petroleum diesel additives and the second are emerging new technologies developed specifically for biodiesel. Traditional petroleum diesel additives can be described as either PP depressants or wax crystalline modifiers. PP depressants were developed to improve the pumpability of crude oil and do not affect nucleation. Instead, these additives inhibit crystalline growth, thereby eliminating agglomeration.[18] They are typically composed of low-molecular-weight copolymers. Wax crystalline modifiers, as the name suggests, are copolymers that disrupt part of the crystallization process to produce a large number of smaller, more compact wax crystals.[19] Studies have been undertaken on both types of additives in both petroleum blends and pure biodiesel. For example, the PP of neat soybean methyl ester was lowered by as much as 6 K.[2,36] Similar improvements in CFPP were achieved, but no noticeable improvement in CP was reported. Researchers have studied the impact of cold flow improvers on soybean biodiesel blends.[36] Four cold flow improvers were tested at 0.1–2% in B80, B90, and B100 blends. Two additives significantly decreased the PPs of soybean biodiesel blends, but all four of the additives had little effect on the CP. A mixture of 0.2% additive, 79.8% biodiesel, and 20% kerosene reduced the PP by 27 °C.

The researchers also investigated the cold flow properties of 100% biodiesel fuels obtained from *Madhuca indica*.[37] The cold flow properties of biodiesel were evaluated with and without PP depressants towards the objective of identifying the pumping and injecting of these biodiesel in Compression ignition (CI) engines under cold climates. The effect of ethanol, kerosene and commercial additives on the cold flow behavior of this biodiesel was studied. A considerable reduction in PP was noticed by using these cold flow improvers. The performance and emission with ethanol-blended mahua biodiesel fuel and ethanol–diesel blended mahua biodiesel fuel have also been studied. Ethanol-blended biodiesel is totally a renewable, viable alternative fuel for improved cold flow behavior.

Knothe *et al.* synthesized fatty diesters *via* the *p*-toluene sulfonic acid-catalyzed esterification of both mono- and bi-functional fatty acids and alcohols.[38] However, CP values for blends with soybean methyl ester were reduced by 1 K at most. Researchers investigated the influence of several additives, which were either synthesized in-house or commercially available, including: Tween-80, dihydroxy fatty acid (DHFA), acrylated polyester prepolymer (APP), palm-based polyol (PPO), a blend of 1 : 1 DHFA and ethylhexanol (DHFAPP), an additive synthesized using DHFA and ethylhexanol (DHFAEH), and castor oil ricinoleate. They reported an average reduction in CP values of 5.5 K, with the largest reduction being 10.5 K for the addition of 1% DHFA and 1% PPO to palm oil methyl ester. They concluded that the effectiveness in particular of the polyhydroxy compounds was due to the interaction between the hydroxy groups of the additives and the samples.

Researchers suggested that the addition of fatty acid moieties whose chain length was similar to the corresponding fatty acid ester, but with protruding polar groups would improve the low-temperature properties. Ozonized vegetable oils of sunflower (SFO), soybean (SBO) and rapeseed (RSO) oil were prepared by passing ozone through the oil in a bubbling bed reactor. Ozonized SFO was added to SFO-, SBO- and RSO-derived biodiesel at levels of 1 or 1.5 wt%. The PPs of these blends of SFO, SBO and RSO were reduced, however the CPs remained unchanged. In the case of similar blends of ozonized SFO with palm oil derived biodiesel, the PP remained unchanged, but the CP was significantly reduced. In further tests, a set of blends was prepared by adding the ozonized oils to biodiesel derived from the same oil. This resulted in the greatest reductions in PP. The PPs of these blends of SFO, SBO and RSO were lowered to −24, −12 and −30 °C, respectively. But ozonolysis on such a large scale would undoubtedly add significantly to the cost of an already marginal commodity like biodiesel.[18]

5.4.2 Removing Solid Precipitates

As already discussed related to the importance of the CSFT, biodiesel fuel manufacturers are making adjustments to ensure their products will pass the CSFT. Factors include choice and quality of lipid feedstocks and purification steps.[26] A decrease in filter-plugging tendency was noted when chilling biodiesel to 4.4–21 °C for selective removal of free steryl glucosides after production.[39,40] The process employs a combination of preparatory steps which include: (a) the use of diatomaceous earths mixed in a slurry with biodiesel or directly in a filter bed; (b) the use of magnesium silicate mixed in a slurry; (c) replacing the water-washing step in the biodiesel production process by mixing with 1.5–2 wt% silica hydrogel under heat and negative pressure (vacuum); and (d) adding deoiled lecithin and subjecting the mixture to a degumming procedure. Each step is followed by controlled filtration to remove solid precipitates. This invention was shown to significantly reduce free steryl glucoside levels and hence improve the low-temperature flow properties without affecting the MAG, diacylglycerol (DAG) or triacylglycerol (TAG) concentrations.

A second process decreases the result of the CSFT by removing solid precipitates after biodiesel production.[41] This step involves cooling biodiesel to 21 °C, mixing with diatomaceous earth to form a slurry and filtering the mixture through a pressure leaf filter followed by a polishing filter to remove any remaining sediments or diatomaceous earth.

5.4.3   Feedstock Modification

Winterization of vegetable oil esters is based on the lowering of the melting points of unsaturated fatty compounds *versus* saturated compounds.[2,42,43] This method removes solids, formed during cooling of the vegetable oil esters, by filtration, leaving a mixture with a higher content of unsaturated fatty esters and thus with lower CP and PP. This procedure can be repeated to further reduce the CPs and PPs. There are basically two techniques used for winterization. One technique involves refrigeration of the oils for a prescribed period at a specific temperature followed by decanting of the remaining liquid. Another, more energy-efficient method, is to allow tanks of oil to stand outside in cold temperatures for extended periods of time. In either case, the fraction that remains molten is separated from the solid-producing oil with improved pour and handling qualities.[18] There are reports of the improvement of low-temperature flow properties with winterization.[2,44,45]

Another method for altering biodiesel feedstocks is to genetically modify the fatty acid profile of oilseeds.[46] It has been suggested that soybean could be modified to produce an oil profile with elevated oleic acid (C18 : 1) and with reduced polyunsaturated and saturated fatty acids. Such an oil would be useful in terms of low-temperature properties in most climates.

5.4.4   Biodiesel Modification

This technique for modifying the low-temperature properties of biodiesel involves crystallization fractionation. Both dry fractionation and solvent fractionation have been applied to biodiesel. Dry fractionation involves crystallization from the melt without the addition of a solvent and is the simplest and least expensive method.[19] This method is used to improve the low-temperature flow properties.[42] Solvent fractionation has advantages over dry fractionation, including reduced crystallization times and improved yields, but suffers from reduced safety and increased cost. Hexane extraction has been employed in a single-step process with a residence time of 3.5–6.5 h for soybean methyl ester.[43] Other researchers have attempted solvent fractionation with various solvents, including methanol, acetone, chloroform, and ethanol, with limited success.[42,47] Fractionation not only introduces a new set of unit operations including solvent addition, crystallization and solvent recovery, but it also results in significantly reduced yields. The loss of 75% of the original material would be unacceptable for any commercial process.

Research into chemically converting biodiesel to triesters has been reported.[48] Researchers synthesized palm oil polyol esters from fractionated palm oil methyl esters and trimethylolpropane using sodium methoxide as a catalyst. PPs of -30 °C were achieved for a product that was greater than 90 wt% triester. But the chemical modification of palm oil based biodiesel was only successful when combined with fractionation to significantly reduce the C16 : 0 content.

Researchers have exploited the π-bonds of unsaturated fatty acids *via* electrophilic addition to produce branched or bulky esters to improve the low-temperature flow properties of biodiesel.[49,50] Cermack *et al.* prepared the estolides of oleic acid and tallow fatty acids, at varying ratios, by treating the fatty acids with perchloric acid at 60 °C under vacuum.[51] After 24 h, 2-ethylhexanol was added, the vacuum restored and the reaction continued for a further 4 h. As the amount of tallow in the starting material was decreased, the PP and CP improved.

Surrogate molecules that is, methyl, ethyl, isopropyl and butyl esters of β-branched fatty acids, having substantially better low-temperature flow properties have also been synthesized.[52]

5.4.5 Blending of Biodiesels

The properties of the various individual fatty esters that comprise biodiesel determine its overall fuel properties, and, in turn, the properties of various fatty esters are determined by their structural features. Blending of biodiesels with different FAME compositions is therefore expected to improve low-temperature flow properties.[27,28]

When palm and jatropha biodiesels or palm and pongamia biodiesels are blended, the blended biodiesel will have improved low-temperature flow properties compared with PBD.[27,28]

PBD, JBD and PoBD were blended with different weight ratios (%) as follows: 100 : 00 : 00, 80 : 20 : 00, 80 : 00 : 20, 60 : 40 : 00, 60 : 20 : 20, 60 : 00 : 40, 40 : 60 : 00, 40 : 40 : 20, 40 : 20 : 40, 40 : 00 : 60, 20 : 80 : 00, 20 : 60 : 20, 20 : 40 : 40, 20 : 20 : 60, 20 : 00 : 80, 00 : 100 : 00, 00 : 80 : 20, 00 : 60 : 40, 00 : 40 : 60,

00 : 20 : 80, 00 : 00 : 100 respectively. The aforementioned 21 biodiesel samples were used for the experiment.

The contents of saturated FAMEs for palm, jatropha, and pongamia biodiesels were 44.4, 21.1, and 16.0% respectively and the contents of unsaturated FAMEs for the three biodiesels were 55.6, 78.9 and 84.0% respectively (Table 5.2).[27,28]

PBD and JBD were blended in weight percentages: 100 : 00 (PBD), 80 : 20 (PJBD-1), 60 : 40 (PJBD-2), 40 : 60 (PJBD-3), 20 : 80 (PJBD-4), and 00 : 100 (JBD).[27,28] Figure 5.14 shows that the high CP of PBD was lowered after blending with JBD, which has a low CP.[28] Figure 5.14 also shows that the PP of blended biodiesel improved significantly and even dropped to 0 °C when JBD was present at 80 wt%.

When PBD was blended with PoBD in weight percentages: 100 : 00 (PBD), 80 : 20 (PPoBD-1), 60 : 40 (PPoBD-2), 40 : 60 (PPoBD-3), 20 : 80 (PPoBD-4), and 00 : 100 (PoBD), the CP of the blended biodiesel improved significantly (Figure 5.15).[28] The PP of blended PBD dropped to below 3 °C when PoBD was present at more than 40 wt% (Figure 5.15). As pure PBD, which has a high CP and PP, could not be used in the winter season, blending with JBD and PoBD, which have flow CPs and PPs, remarkably improved the low-temperature flow properties of PBD and therefore, the use of edible palm oil can be minimized.

JBD and PoBD were blended in weight percentages: 100 : 00 (JBD), 80 : 20 (JPoBD-1), 60 : 40 (JPoBD-2), 40 : 60 (JPoBD-3), 20 : 80 (JPoBD-4), and 00 : 100 (PoBD). By adding PoBD, which has a relatively lower CP, to JBD, the CP of JBD blends improved (Figure 5.16).[28] Figure 5.16 also shows that the PP of blended JBD improved when PoBD was present at 80 wt%.

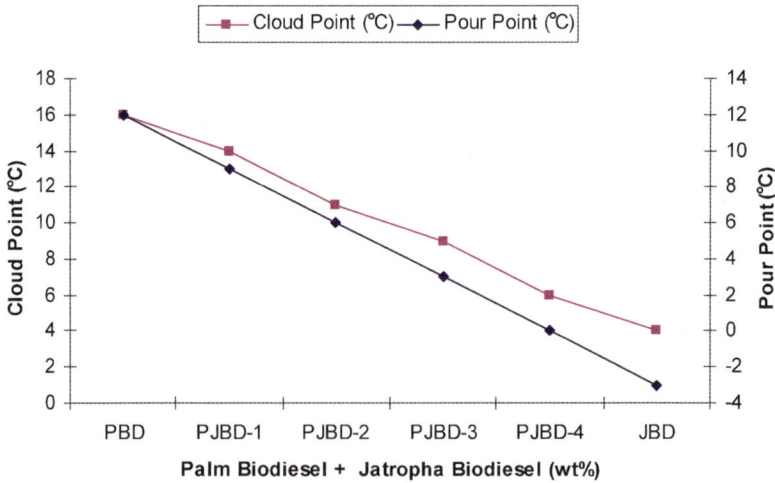

Figure 5.14   CPs and PPs of palm biodiesel blended with jatropha biodiesel.

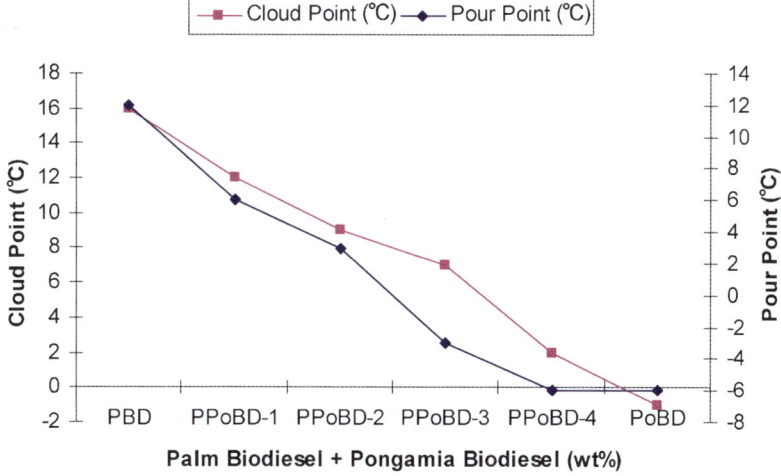

Figure 5.15 CPs and PPs of palm biodiesel blended with pongamia biodiesel.

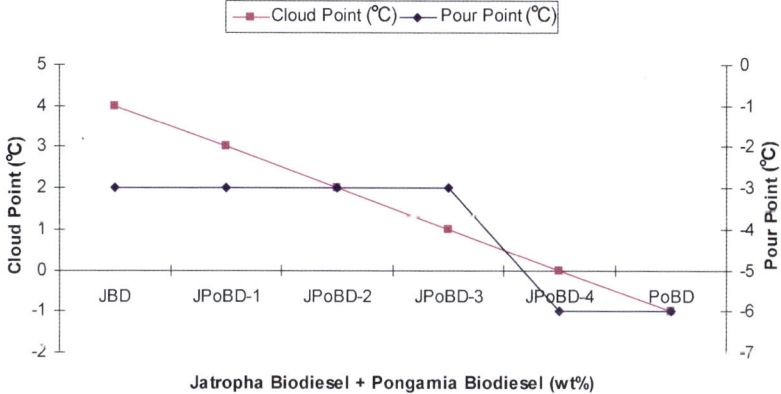

Figure 5.16 CPs and PPs of jatropha biodiesel blended with pongamia biodiesel.

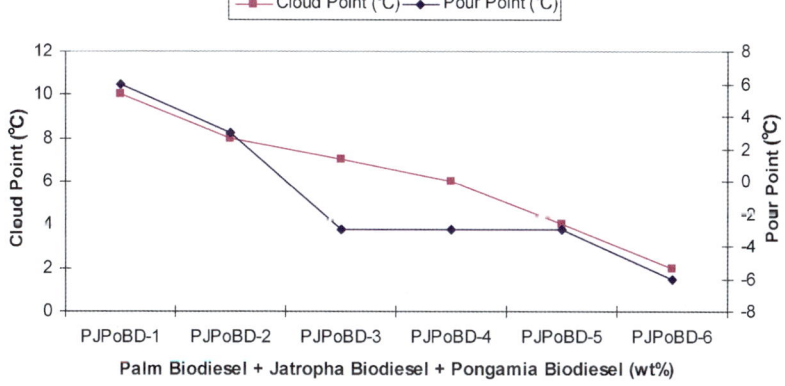

Figure 5.17 CPs and PPs of palm biodiesel blended with jatropha and pongamia biodiesel.

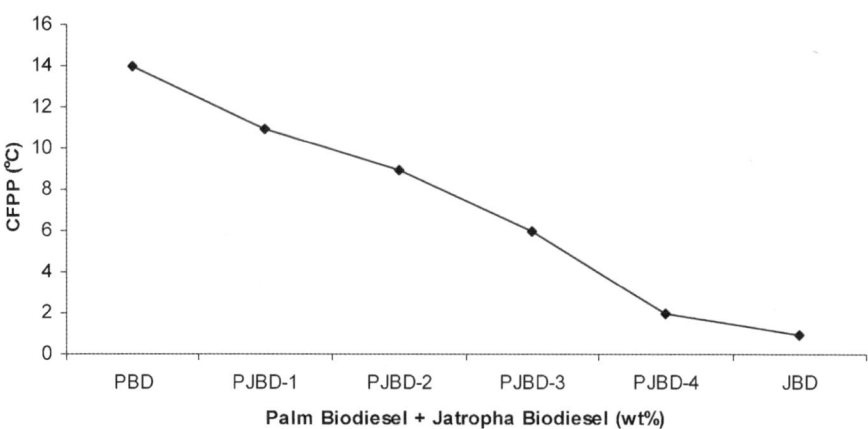

Figure 5.18   The CFPP of palm biodiesel blended with jatropha biodiesel.

PBD, JBD and PoBD were blended in weight percentages: 60 : 20 : 20 (PJPoBD-1), 40 : 40 : 20 (PJPoBD-2), 40 : 20 : 40 (PJPoBD-3), 20 : 60 : 20 (PJPoBD-4), 20 : 40 : 40 (PJPoBD-5), and 20 : 20 : 60 (PJPoBD-6). Figure 5.17 shows that the CP of PBD decreased after blending with JBD and PoBD.[28] Blending PBD with JBD and PoBD, which have low PPs remarkably improved the PP of PBD.

Further tests were done to investigate the blending effect of biodiesels on the CFPP.[27] PBD and JBD were blended in weight percentages: 100 : 00 (PBD), 80 : 20 (PJBD-1), 60 : 40 (PJBD-2), 40 : 60 (PJBD-3), 20 : 80 (PJBD-4), and 00 : 100 (JBD). Figure 5.18 shows that the CFPP of PBD improved significantly when JBD was blended with it in varying weight ratios.

When PBD was blended with PoBD in weight percentages: 100 : 00 (PBD), 80 : 20 (PPoBD-1), 60 : 40 (PPoBD-2), 40 : 60 (PPoBD-3), 20 : 80 (PPoBD-4), and 00 : 100 (PoBD), the CFPP of blended PBD dropped to below 1 °C when PoBD was present at 60 wt% (Figure 5.19).[27]

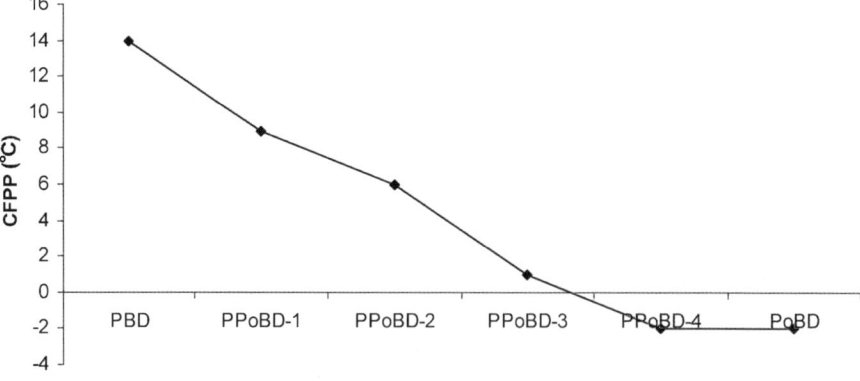

Figure 5.19   The CFPP of palm biodiesel blended with pongamia biodiesel.

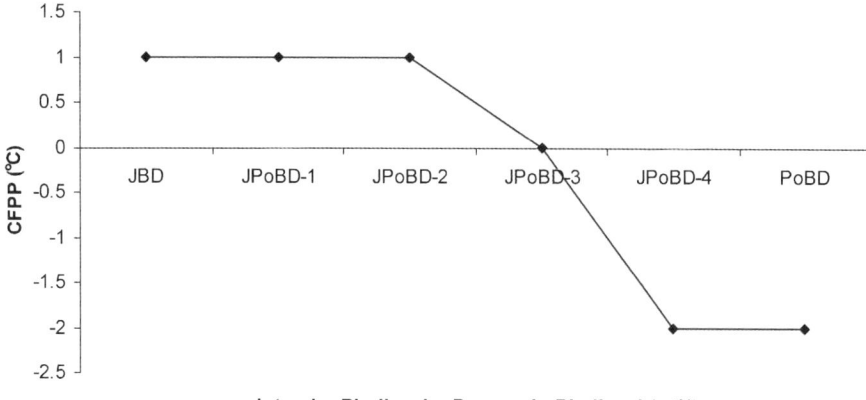

Figure 5.20   The CFPP of jatropha biodiesel blended with pongamia biodiesel.

JBD and PoBD were blended in weight percentages: 100 : 00 (JBD), 80 : 20 (JPoBD-1), 60 : 40 (JPoBD-2), 40 : 60 (JPoBD-3), 20 : 80 (JPoBD-4), and 00 : 100 (PoBD). Figure 5.20 shows that the CFPP of JBD was improved when PoBD was blended with it.[27]

PBD, JBD and PoBD were blended in weight percentages: 60 : 20 : 20 (PJPoBD-1), 40 : 40 : 20 (PJPoBD-2), 40 : 20 : 40 (PJPoBD-3), 20 : 60 : 20 (PJPoBD-4), 20 : 40 : 40 (PJPoBD-5), and 20 : 20 : 60 (PJPoBD-6).[27] Blending PBD with JBD and PoBD, which have low CFPPs remarkably improved the CFPP of PBD (Figure 5.21).[27]

Researchers blended palm, rapeseed and soybean biodiesels and found that the high CFPP of PBD was improved after blending with the rapeseed and soybean biodiesels.[29] It has been reported that the very poor low-temperature

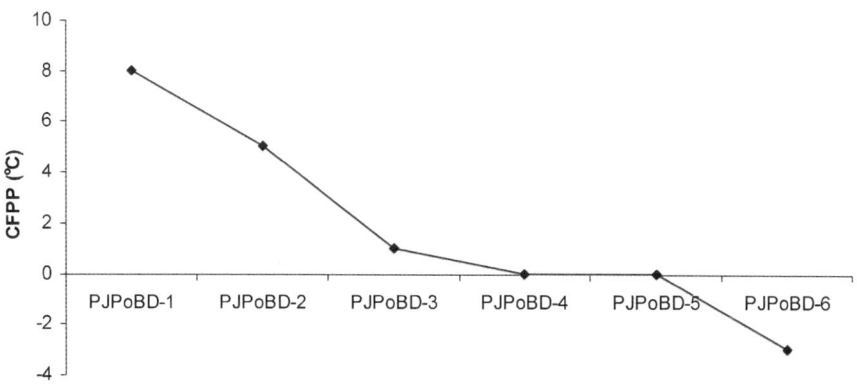

Figure 5.21   The CFPP of palm biodiesel blended with jatropha and pongamia biodiesels.

performance of palm-based biodiesel can be improved by blending with jatropha-based fuel.[53]

Therefore, to improve poor biodiesel low-temperature flow properties, blending of two or more biodiesels is a simple but effective method.

References

1. R. O. Dunn and M. O. Bagby, *J. Am. Oil Chem. Soc.*, 1995, **72**, 895.
2. R. O. Dunn, M. W. Shockley and M. O. Bagby, *J. Am. Oil Chem. Soc.*, 1996, **73**, 1719.
3. H. Imahara, E. Minami and S. Saka, *Fuel*, 2006, **85**, 1666.
4. J. A. P. Coutinho, F. Mirante, J. C. Ribeiro, J. M. Sansot and J. L. Daridon, *Fuel*, 2002, **81**, 963.
5. G. Knothe, *Fuel Process. Technol.*, 2005, **86**, 1059.
6. Conservation of clean air and water in Europe special task force, FE/STF-24, Report No. 9/09, CONCAWE, Brussels, 2009.
7. H. Y. Tang, R. C. De Guzman, S. O. Salley and K. Y. S. Ng, *J. Am. Oil Chem. Soc.*, 2008, **85**, 1173.
8. N. Ohshio, K. Saito, S. Kobayashi and S. Tanaka, SAE Paper No. 2008-01-2505SAE International, Warrendale, 2008.
9. H. Y. Tang, S. O. Salley and K. Y. S. Ng, *Fuel*, 2008, **87**, 3006.
10. Imperial Oil, Report No. R497-2009, Natural Resources Canada, Ottawa, Canada, 2009.
11. Imperial Oil, Report No. R498-2009, Natural Resources Canada, 2009.
12. A. Serdari, E. Lois and S. Stournas, *Ind. Eng. Chem. Res.*, 1999, **38**, 3543.
13. S. Stournas, E. Lois and A. Serdari, *J. Am. Oil Chem. Soc.*, 1995, **72**, 433.
14. A. A. Refaat, *Int. J. Environ. Sci. Technol.*, 2009, **6**, 677.
15. T. Foglia, L. Nelson, R. Dunn and W. Marmer, *J. Am. Oil Chem. Soc.*, 1997, **74**, 951.
16. J. Rodrigues, F. Cardoso, E. Lachter, L. Estevao, E. Lima and R. Nascimento, *J. Am. Oil Chem. Soc.*, 2006, **83**, 353.
17. G. L. Hasenhuetti, in *Kirk–Othmer Encyclopedia of Chemical Technology*, J. I. Kroschwitz and A. Seidel, John Wiley & Sons, New York, 2005, p. 816.
18. P. C. Smith, Y. Ngothai, Q. D. Nguyen and B. K. O'Neill, *Renewable Energy*, 2010, **35**, 1145.
19. G. Knothe, J. Krahl and J. V. Gerpen, *The Biodiesel Handbook*, AOCS Press, Champaign, 2004.
20. R. Kotrba, *Biodiesel Mag.*, 2006, **3**, 42.
21. I. Lee, L. M. Pfalzgraf, G. B. Poppe, E. Powers and T. Haines, *Biodiesel Mag.*, 2007, **4**, 105.
22. V. Van Hoed, N. Zyaykina, W. De Greyt, J. Maes, R. Verhé and K. Demeestere, *J. Am. Oil Chem. Soc.*, 2008, **85**, 701.

23. J. H. V. Gerpen, E. G. Hammond, L. A. Johnson, S. J. Marley, L. Yu, I. Lee and A. Monyem, Iowa Soybean Promotion Board, Iowa state university, Ames, Iowa, USA, 1996.
24. L. Pfalzgraf, I. Lee, J. Foster and G. Poppe, *Biorenewable Resources*, AOCS Press, Champaign, 2007, no. 4, p.17.
25. C. Selvidge, S. Blumenshine, K. Campbell, C. Dowell and J. Stolis, presented at the 10th International Conference on Stability, Handling and Use of Liquid Fuels, Atlanta, 2007.
26. R. Kotrba, *Biodiesel Mag.*, 2009, **6**, 38.
27. A. Sarin, R. Arora, N. P. Singh, R. Sarin, R. K. Malhotra and S. Sarin, *Energy Fuels*, 2010, **24**, 1996.
28. A. Sarin, R. Arora, N. P. Singh, R. Sarin, R. K. Malhotra and K. Kundu, *Energy*, 2009, **34**, 2016.
29. J. Park, D. Kim, J. Lee, S. Park, Y. Kim and J. Lee, *Bioresour. Technol.*, 2008, **99**, 1196.
30. C. R. Krishna, K. Thomassen, C. Brown, T. A. Butcher, M. Anjom and D. Mahajan, *Ind. Eng. Chem. Res.*, 2007, **46**, 8846.
31. J. C. A. Lopes, L. Boros, M. A. Krahenbuhl, A. J. A. Meirelles, J. L. Daridon, J. Pauly, I. M. Marrucho and J. A. P. Coutinho, *Energy Fuels*, 2008, **22**, 747.
32. A. Pradhan and D. S. Shrestha, presented at the ASABE Annual International Meeting, St. Joseph, Paper No. 076090, 2007.
33. N. Pasadakis, S. Sourligas and C. Foteinopoulos, *Fuel*, 2006, **86**, 1131.
34. R. M. Joshi and M. J. Pegg, *Fuel*, 2007, **86**, 143.
35. P. Benjumea, J. Agudelo and A. Agudelo, *Fuel*, 2008, **87**, 2069.
36. C. Chiu, L. G. Schumacher and G. L. Suppes, *Biomass Bioenergy*, 2004, **27**, 485.
37. P. V. Bhale, N. V. Deshpande and S. V Thombre, *Renewable Energy*, 2009, **34**, 794.
38. G. Knothe, R. O. Dunn, M. W. Shockley, and M. O. Bagby, *J. Am. Oil Chem. Soc.*, 2000, **77**, 865.
39. I. Lee, J. L. Mayfield, L. M. Pfalzgraf, L. Solheim and S. Bloomer, *US Pat.*, 0151146, 2007.
40. I. Lee, L. M. Pfalzgraf, L. Solheim and S. Bloomer, *EU Pat.*, 969090, 2008.
41. M. F. Danzer, T. L. Ely, S. A. Kingery, W. W. McCalley, W. M. McDonald, J. Mostek and M. L. Schultes, *US Pat.*, 0175091, 2007.
42. I. Lee, L. A. Johnson and E. G. Hammond, *J. Am. Oil Chem. Soc.*, 1996, **73**, 631.
43. R. O. Dunn, M. W. Shockley, M. O. Bagby, SAE Paper No. 971682SAE International, Warrendale, 1997.
44. M. E. G. Gómez, R. Howard-Hildige, J. J. Leahy and B. Rice, *Fuel*, 2002, **81**, 33.
45. Á. Pérez, A. Casas, C. M. Fernández, M. J. Ramos and L. Rodríguez, *Bioresour. Technol.*, 2010, **101**, 7375.

46. J. Duffield, H. Shapouri, M. Graboski, R. McCormick and R. Wilson, USDA, Society of automotive engineers, Warrendale, USA, 1998.
47. M. A. Hanna, Y. Ali, S. L. Cuppett and D. Zheng, *J. Am. Oil Chem. Soc.*, 1996, **73**, 759.
48. R. Yunus, A. Fakhru'l-Razi, T. L. Ooi, R. Omar and A. Idris, *Ind. Eng. Chem. Res.*, 2005, **44**, 8178.
49. B. R. Moser and S. Z. Erhan, *J. Am. Oil Chem. Soc.*, 2006, **83**, 959.
50. B. R. Moser and S. Z. Erhan, *Fuel*, 2008, **87**, 2253.
51. S. C. Cermack, A. L. Skender, A. B. Deppe and T. A. Isbell, *J. Am. Oil Chem. Soc.*, 2007, **84**, 449.
52. R. Sarin, R. Kumar, B. Srivastav, S. K. Puri, D. K. Tuli, R. K. Malhotra and A. Kumar, *Bioresour. Technol.*, 2009, **100**, 3022.
53. R. Sarin, M. Sharma, S. Sinharay and R. K. Malhotra, *Fuel*, 2007, **86**, 1365.

CHAPTER 6

Dependence of Other Properties of Biodiesel on Fatty Acid Methyl Ester Composition and Other Factors

6.1 Introduction

As discussed earlier, in order for biodiesel to be used commercially as a fuel, the finished biodiesel must be analyzed using sophisticated analytical equipment to ensure it meets international standards. Besides oxidation stability and low-temperature flow properties there are many other important properties like viscosity, cetane number, flash point and others which may depend upon the fatty acid methyl ester (FAME) composition or may be interdependent. This chapter deals with these properties.

6.2 Prediction of Viscosity

Viscosity is a measure of resistance to flow of a liquid due to the internal friction of one part of a fluid moving over another. The affects of high viscosity are discussed in Chapter 3. The viscosity of biodiesel is typically higher than that of petroleum diesel—often by a factor of two. The viscosity of biodiesel blends increases as the blend level increases. The viscosity of straight vegetable oil is much higher yet, and is the main reason why such oils are unacceptable as diesel blendstocks and viscosity is reduced by transesterification. Viscosity is greatly affected by temperature.[1] Therefore, many of the problems resulting

Biodiesel: Production and Properties
Amit Sarin
© Amit Sarin 2012
Published by the Royal Society of Chemistry, www.rsc.org

from high viscosity are most noticeable under low ambient temperature and cold-start engine conditions. A study has shown that as the temperature is reduced, the distribution of B100 fuel among individual injectors within an injector assembly becomes very unequal and this, in turn, could lead to engine performance and emissions problems.[2] The viscosity of individual FAME molecules is known to increase with fatty acid carbon number.[3,4]

The Grunberg–Nissan equation has been reported to be the most suitable equation for computing the viscosity of liquid mixtures.[5-7] This equation was developed primarily for binary mixtures and works best with non-associated liquids. The Grunberg–Nissan equation is:

$$\ln V_m = \sum_i x_i \ln V_i + \sum_{i \neq j}^{n} \sum x_i x_j G_{ij} \tag{1}$$

where V_m is the mean viscosity of mixture (Pa s), V_i the viscosity of the pure ith component (Pa s), x_i and x_j the mole fractions of the ith and jth components, G_{ij} the interaction parameter (Pa s), and n the number of components.

Biodiesel fuels are non-associated liquids that are comprised of mixtures of fatty acid esters whose chemical structures are similar. Due to this similarity, the components in a mixture should not interact with each other and thus should behave in a similar manner as an individual component. It was therefore assumed that the interaction parameter in eqn (1) would be small and thus could be neglected. Also, the mass fraction was used in preference to the mole fraction in eqn (1) to conform to the mass unit that is implicit in the units for viscosity used in this study. With these two modifications, eqn (2) was used to predict the viscosity of biodiesel fuels based on their fatty acid composition.

$$\ln V_m = \sum_{i=1}^{n} y_i \ln V_i \tag{2}$$

where y_i is the mass fraction.

Using experiments, researchers have verified this equation for predicting the viscosities of biodiesel fuels from knowledge of their fatty acid composition.[8] The applicability of a logarithmic mixture equation was verified using controlled mixtures of standard fatty acid esters and natural biodiesels. Several binary, ternary and quaternary mixtures of fatty acid ethyl ester (FAEE) GC standards were formulated. Their viscosities were predicted from their component values and were within 3.7% of their measured values. The fatty acid compositions of six typical oils were simulated by mixing FAME standards in appropriate amounts, and the viscosities of these mixtures were predicted to within 2.1% of their measured values. Five biodiesel types were produced from natural oils and the logarithmic equation was applied to predict their viscosities. An average prediction error of 3% was obtained for these samples. The viscosities of 15 biodiesel types were then predicted based on their fatty acid composition, as published in the literature, and were found to vary by as much as 100%. This is most likely a principal contributing factor to the variation in performance of some biodiesel fuel types. The viscosity of biodiesel fuels reduces considerably with increasing unsaturation. Contamination with small amounts of glycerides significantly affects the viscosity of biodiesel. Eqns (3) and (4) give the fitted polynomials for saturated methyl esters (MEs) and ethyl esters (EEs) respectively:

$$V_{\text{ME-sat}} = 1.05E - 4M^2 = 0 : 0242M + 2.15(s = 0 : 0145) \tag{3}$$

$$V_{\text{EE-sat}} = 1.16E - 4M^2 - 0 : 0264M + 2.28(s = 0 : 0182) \tag{4}$$

where V is the viscosity (mPa s), M the molecular weight (g mol^{-1}) and s the standard error. The viscosity trend for unsaturated esters at 40 °C showed a sharp deviation from the trend of the saturated esters when C18 : 0 became C18 : 1. As the number of double bonds increased, there was a non-linear decrease in viscosity, with a 21% difference between C18 : 0 and C18 : 1 (based on C18 : 0), an 18% difference between C18 : 1 and C18 : 2 (based on C18 : 1), and a 13% difference between C18 : 2 and C18 : 3 (based on C18 : 2). The trend for unsaturated C18 esters also correlated well with the second-order polynomial functions given in eqns (5) and (6) for MEs and EEs, respectively.

$$V_{\text{ME-unsat}} - C18 = 0.153NDB^2 - 1.15NDB + 4 : 73(s = 0.0112) \tag{5}$$

$$V_{\text{EE-unsat}} - C18 = 0 : 147NDB^2 - 1.09NDB + 4 : 82(s = 0 : 000) \tag{6}$$

where NDB is the number of double bonds in the C18 chain.

In another work, the kinematic viscosity of numerous fatty compounds as well as components of petrodiesel were determined at 40 °C.[9] The objective was to obtain a database on kinematic viscosity under identical conditions that can be used to define the influence of compound structure on kinematic viscosity. Kinematic viscosity increases with chain length of either the fatty acid or alcohol moiety in a fatty ester or in an aliphatic hydrocarbon. The increase in kinematic viscosity over a certain number of carbons is smaller in aliphatic hydrocarbons than in fatty compounds. It was found that the viscosity of unsaturated fatty compounds strongly depends on the nature and number of double bonds, with the double-bond position affecting viscosity less. Terminal double bonds in aliphatic hydrocarbons have a comparatively small viscosity-reducing effect. Branching in the alcohol moiety does not significantly affect viscosity compared to straight-chain analogues. Free fatty acids or compounds with hydroxy groups possess significantly higher viscosities. The viscosity range of fatty compounds is greater than that of the various hydrocarbons comprising petrodiesel. The effect of dibenzothiophene, a sulfur-containing compound found in petrodiesel fuel, on the viscosity of toluene is less than that of fatty esters or long-chain aliphatic hydrocarbons. The influence of the nature of oxygenated moieties on kinematic viscosity in compounds with 10 carbons and varying oxygenated moieties was also investigated. A reversal in the effect on viscosity of the carboxylic acid moiety *vs.* the alcohol moiety was noted for C10 compounds compared to unsaturated C18 compounds. The sequence of influence on kinematic viscosity of oxygenated moieties is COOH $\approx$ C–OH $>$ OCOOCH$_3$ $\approx$ C=O$>$ C–O–C $>$ no oxygen.

An empirical equation for calculating the dynamic viscosity of biodiesel derived from ethyl esters of fish oil, No. 2 diesel fuel, and their blends as a function of both temperature and blend has been developed.[10] The dynamic viscosities of biodiesel derived from ethyl esters of fish oil, No. 2 diesel fuel, and their blends were measured from 298 K down to their respective PPs. Blends of B80 (80 vol% biodiesel–20 vol% No. 2 diesel), B60, B40 and B20 were investigated. The equation is:

$$\ln V = -2.4343 + 216.66/T + 293523/T^2 \tag{7}$$

where V is the dynamic viscosity (mPa S) and T is the temperature (K). Eqn (7) gives a standard deviation of 0.0223 and a correlation coefficient (R^2) of 100%.

In 2007, to predict the viscosity of any given biodiesel fuel (FAME mixture), a novel topological index based on the distance matrix and adjacent matrix of the molecular structure was proposed.[11] Viscosity is a physical property that is closely related to the molecular structure, length of the fatty acid hydrocarbon chain and the number of unsaturated bonds. The longer the fatty acid hydrocarbon chain, the higher the viscosity of the FAME; on the other hand, the greater the number of unsaturated bonds, the lower the viscosity of the FAME. Therefore, a molecular structural topological index can effectively reflect viscosity through the quantitative description of the characteristics of molecular structure. The new topological index can reflect information on the molecular structure of the FAME, such as the size of the molecule, the number of unsaturated bonds and the branch degree. Combined with the modified Grunberg–Nissan or Hind equation, the topological index values of the FAME mixture were calculated. Then, by relating the topological index values of the FAME mixtures with their viscosities, two linear regression equations were obtained. Using these regression equations, the viscosity of biodiesel fuels were predicted. The results show that the modified Grunberg–Nissan equation has a higher precision of prediction than the Hind regression equation. The correlation between V and the mean mixture topological index (χ_m) of biodiesel is expressed in regression eqns (8) and (9):

$$V = 5.96249\chi_m - 10 : 08814 \tag{8}$$

$$V = \chi_m + 0.038 \tag{9}$$

Using eqn (8), the viscosities of the methyl esters of peanut oil, rapeseed oil, canola oil, and coconut oil, palm oil and soybean oil were predicted. The predicted viscosities of six biodiesel fuels are in good agreement with the reported data. The highest relative error of 9.2% occurred with the methyl ester of soybean oil, and the methyl ester of coconut oil had the lowest relative error of 0.46%. Thus, the assumption of a negligible interaction parameter, the use of the measured mass fraction of the components, and the use of topological index values provides acceptable results. The comparatively high error for the methyl ester of soybean oil could have been due to the methyl ester of soybean oil having a high percentage of C18 : 2 and C18 : 3 (62.4% in total mass), while the methyl ester of coconut oil only has a low percentage of C18 : 2 and C18 : 3 (1.7% in total mass). Using eqn (9), the viscosities of the methyl esters of peanut oil, rapeseed oil, canola oil, and coconut oil, palm oil and soybean oil were also predicted, and then compared to the data published in the literature. The predicted viscosities are not in good agreement with the reported data. The highest relative error of 22.61% occurred with the methyl ester of canola oil. Although the methyl ester of peanut oil had the lowest relative error of 0.04%, relative errors of the methyl esters of rapeseed oil and soybean oil are still high at 19.32% and 11.31%, respectively.

An index termed here the low-temperature viscosity ratio (LTVR) using data at 0 °C and 40 °C (by dividing the viscosity value at 0 °C by the value at 40 °C) was used to evaluate individual compounds but also mixtures by their low-temperature viscosity behavior.[1] The compounds tested included a variety of saturated, monounsaturated, di-unsaturated and tri-unsaturated fatty esters, methyl ricinoleate—in which the OH group leads to a significant increase in viscosity—as well as triolein and some fatty alcohols and alkanes. Esters of oleic acid have the highest viscosity of all biodiesel components that are liquids at low temperatures. The behavior of blends of biodiesel and some fatty esters with a low-sulfur diesel fuel was also investigated The low temperature viscosity behavior of biodiesel, its components, related fatty materials and blends of fatty compounds as well as blends with petrodiesel were investigated. The blends showed a behavior closer to that of petrodiesel than of biodiesel or its neat components. Esters with shorter fatty acid chains but longer alcohol moieties display lower viscosities than esters with longer fatty acid chains and shorter alcohol moieties. Saturated esters with high melting points have only a small influence on kinematic viscosity at lower temperatures and at concentrations observed in many common vegetable oils. The result can be used for assessing fatty esters in terms of enriching them in biodiesel for the improvement of fuel properties and the data can be used for predicting or verifying the viscosity of as yet non-investigated compounds as well as of mixtures.

The properties of various potential major components of biodiesel have been examined and compared by researchers.[12] For example, while methyl oleate has been suggested as such a major component, methyl palmitoleate has advantages compared to methyl oleate, especially with regards to low-temperature properties. Other materials that have been examined in this context are short-chain (C8–C10) saturated esters, with only C10 esters appearing to be suitable. It has also been suggested that to obtain biodiesel fuel with favorable properties, it is advantageous for the fuel to consist of only one major component in as high a concentration as possible; however, mixtures of components with advantageous properties as described here may also be acceptable. The kinematic viscosity of methyl oleate increases from 4.51 mm^2 s^{-1} at 40 °C to 21.33 mm^2 s^{-1} at -10 °C, with the latter value approaching the kinematic viscosity of some vegetable oils at 40 °C. Palmitoleic acid occurs in small amounts in various vegetable oils, animal fats, and fish oils. It offers advantages in terms of kinematic viscosity: in particular the difference at low temperatures may prove to be significant.

Mathematical relationships between the higher heating value (HHV), the viscosity (V), the density (D) and the flash point (FP) of various biodiesel fuels have been estimated.[13] The relations between V and D are, for vegetable oils:

$$V = -0 : 7328D + 938 : 57 \tag{10}$$

and for biodiesels:

$$V = -16 : 155D + 930 : 78 \tag{11}$$

with coefficient of regression (R) values of 0.9398, and 0.9902, respectively. The relation between V and FP for biodiesels is

$$V = 22 : 981FP + 346 : 79 \tag{12}$$

with an R value of 0.9819.

In another study, two different commercially available diesel fuels were blended with the biodiesels produced from six different vegetable oils (sunflower, canola, soybean, cotton seed, corn oil and waste palm oil) and the generalized equations for predicting the viscosities for the blends were given.[14] The blends (B2, B5, B10, B20, B50 and B75) were prepared on a volume basis. For all blends, it was found that there is excellent agreement between the measured and estimated values of the viscosities. The general form of the equation of viscosity as a function of biodiesel fraction is given by

$$V = Ax^2 + Bx + C \tag{13}$$

where A, B, and C are coefficients and x is the biodiesel fraction. The general form of the equation as a function of biodiesel and diesel fuel concentration is given in eqn (14). Although it is necessary to use mass fractions, it was found that the estimated values were closer to the measured values when using volume fractions instead. Therefore, in this study, volume fractions were used instead of mass fractions for predicting viscosity of the blends:

$$\log V_B = m_1 \log V_1 + m_2 \log V_2, \qquad (14)$$

where V_B is the kinematic viscosity of the blend (mm^2 s^{-1}), V_1 and m_1 are the kinematic viscosities of component 1 and its fraction, and V_2 and m_2 are the kinematic viscosities of component 2 and its fraction. In the viscosity estimation, the maximum absolute error values between the measured and calculated values are 1.58 and 1.48% for biodiesel–Normal diesel (N.D) fuel blends and 0.78 and 2.25% for biodiesel–shell extra (S.E) fuel blends, respectively.

Another work reports the use of near infrared (NIR) spectroscopy to determine the kinematic viscosity at 40 °C.[15] Principal component analysis was used to perform a qualitative analysis of the spectra and partial least squares regression to develop the calibration models between analytical and spectral data. The results support the notion that near-infrared spectroscopy, in combination with multivariate calibration, is a promising technique to be applied to biodiesel quality control, in both laboratory- and industrial-scale samples. It was possible to obtain a calibration range for the kinematic viscosity from 3.6 to 4.9 mm^2 s^{-1}.

The synthesis of surrogate molecules is particularly useful for generating insight into structure–activity relationships, understanding processes and improving biodiesel performance (Figure 6.1).[16] In order to improve upon the physicochemical properties of biodiesel, methyl, ethyl, isopropyl and *n*-butyl esters of β-branched fatty acids were synthesized, all initiating from β-branched alcohols. β-Branched alcohols upon oxidation gave corresponding acids, which were converted to their esters. The synthesized surrogate esters have viscosities in the range 4.2–4.6 cSt at 40 °C, meeting international diesel and biodiesel standards.

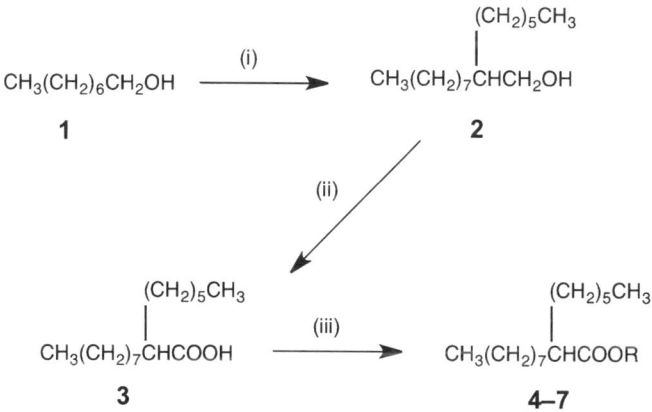

R = methyl (4); ethyl (5); isopropyl (6); butyl (7)

Figure 6.1   The synthesis of biodiesel surrogate molecules.

The viscosities of the synthesized methyl, ethyl, isopropyl and butyl esters are 4.32, 4.38, 4.51 and 4.47 cSt, which meet all of the specifications recommended for diesel and biodiesel. However, a slight increase in viscosity was observed with increasing alkyl chain length. The marginal increase in viscosity, despite the increase in molecular weight, can be attributed to an increase in ester chain length at the O-position, from methyl to ethyl to butyl, without any change in the parent carbon chain.

The purpose of another work was to develop a method for predicting temperature-dependent viscosities of biodiesel based on fatty acid ester composition.[17] Among the available methods, eqns (15) and (16), have been widely used in the oil industry to describe the viscosity–temperature relationship.[18,19]

$$\ln V = A + B/(T + C) \tag{15}$$

$$\log V = B(1/T - 1/T_0) \tag{16}$$

In the above equations, V (mPa s) represents the dynamic viscosity of the liquid at temperature T (K). In eqn (15), A, B, and C are correlation parameters, which can be determined from viscosity measurements at three or more temperatures. For eqn (16), B and T_0 are determined by group contributions on the basis of empirical data. Both methods were extensively studied and reported for pure components but had limited application for fatty acid esters or biodiesel fuels. Considering that experimental data of viscosities of pure fatty acid esters are available, the Vogel equation [eqn (15)] was regarded as both more accurate and simpler to use than eqn (16). Therefore, the Vogel equation with some published viscosity data was employed in this study of biodiesel viscosity prediction. Prediction errors at 25 °C were less than 2.5% for 22 mixtures of fatty acid ethyl esters. Compared with experimentally measured viscosities at 20–100 °C, predicted viscosities of soybean oil and yellow grease methyl esters were within 3%. For coconut, palm and canola oil methyl esters, maximum errors were underestimations at approximately 7%. The maximum average absolute deviation of correlated dynamic viscosities for the methyl esters was 0.77%, indicating that the Vogel equation closely represents the relationship between the dynamic viscosity and temperature of methyl esters. A second-order polynomial correlation was also extrapolated to estimate the dynamic viscosities of these FAMEs, based on the dynamic viscosity calculated for C8 : 0 to C18 : 0. For unsaturated FAMEs, the kinematic viscosities for C22 : 1 were available, but the densities were not known.[20] In addition, the viscosities of C20 : 1 were not available. Therefore, the viscosities of C20 : 1 and C22 : 1 were estimated based on the assumption that the increase of one double bond would have a similar effect on the viscosity among C18 : 1/C18 : 0, C20 : 1/C20 : 0 and C22 : 1/C22 : 1. The procedure for calculating the viscosity of C20 : 1 and C22 : 1 was, first, to determine the relative difference of dynamic viscosity (V_d) between C18 : 1

($V_{18:0}$) and C18 : 0 ($V_{18:1}$) at a given temperature with the following equation:

$$V_d = (V_{18:1} - V_{18:0})/V_{18:0} \tag{17}$$

Then, this difference was applied to C16 : 1/C16 : 0, C20 : 1/C20 : 0 and C22 : 1/C22 : 0 at the same temperatures to calculate the viscosities of C16 : 1, C20 : 1 and C22 : 1 from C16 : 0, C20 : 0 and C22 : 0, respectively. In order to determine the effect of the viscosity estimations for C16 : 1, C20 : 1 and C22 : 1 on biodiesel viscosity predictions, V_d was multiplied by -2, -1, 0, and 2, respectively, in the sensitivity study.

A number of samples of rapeseed oil ethyl ester were prepared by alkaline-catalyzed transesterification under various conditions and the concentrations of the key impurities for biodiesel quality (the concentrations of monoglycerides, diglycerides, triglycerides, free glycerol, ethanol, free fatty acids, water) and some qualitative parameters (flash point, carbon residue, kinematic viscosity at 40 °C) were determined and the relationships between them studied.[21] The relationships were characterized by linear or non-linear statistical models. These models enable a better understanding of the significance of the qualitative parameters and estimate them from the concentrations of impurities. In the case of the viscosity, all investigated impurities were able to influence the kinematic viscosity of the ester phase. A simple linear model with the insignificant influence of potassium ion content and the summation of the glyceride concentration was used for reasons of collinearity. The result of regression analysis is shown in eqn (18):

$$V = 4 : 55 + 11.1W_g + 58.9W_{FFA} - 21.4W_{ET} + 33.4W_{FG} \quad (R^2 = 0 : 952) \tag{18}$$

where W_g = weight of glycerol, W_{FFA} = weight of free fatty acids, W_{ET} = weight of ethanol and W_{FG} = weight of free glycerol.

The resulting model shows a significant increasing influence of glycerides, free glycerol and free fatty acids and a decreasing influence of ethanol. Thus, the viscosity of biodiesel increases because of impurities with higher viscosities and *vice versa*. In contrast, the influence of water content was evaluated as insignificant.

One more investigation has been focused towards the effect of temperature on density and viscosity for a variety of biodiesels and also to develop a correlation between density and viscosity for these biodiesels.[22] The second objective was to investigate and quantify the effects of density and viscosity of biodiesels and their blends on various components of the engine fuel supply system such as the fuel pump, fuel filters and fuel injector. To achieve the first objective, the densities and viscosities of rapeseed oil biodiesel, corn oil biodiesel and waste oil biodiesel blends (B0, B5, B10, B20, B50, B75, and B100) were tested at different temperatures using EN ISO 3675:1998 and EN ISO 3104:1996 standards. The empirical correlation of the kinematic viscosity and temperature is described by eqn (19) for the experimental results obtained in this study.

$$\ln V = -0:0219T + 9:12 \tag{19}$$

where T is the temperature in K, and V is the kinematic viscosity in mm^2 s^{-1}. Furthermore, the density and kinematic viscosity of biodiesel were correlated and an equation was developed relating viscosity as a function of density as shown below in eqn (20). This equation can be used to estimate the kinematic viscosity of biodiesel from a known density value. The values from the model and the experimental values have a maximum absolute error of 0.37.

$$\ln V = 0:0357T = 29:02 \tag{20}$$

where T is the temperature in K and V is the kinematic viscosity in mm^2 s^{-1} of the biodiesel at a given temperature. The empirical equations and the measured data are closely matched with $R^2 = 0.993$ for the density model and 0.999 for the kinematic viscosity model. Furthermore an attempt has been made to develop a correlation to predict the viscosity of a biodiesel blend at a given temperature. For this, a new equation [eqn (21)] has been developed:

$$\ln V_{mix} = (-0:0012T + 0:8456)x - 0:0234T + 8:64 \tag{21}$$

where V_{mix} is the viscosity of the biodiesel, x is the volume fraction of the biodiesel and T is the temperature (K). The modified mixing equation has a maximum absolute error of 0.50. The kinematic viscosity prediction models of biodiesel described by eqns (19)–(21) have been used to determine numerically the kinematic viscosity at various temperatures, density and biodiesel fraction values. These models can be used in the design and investigation of fuel supply systems (fuel pump, fuel filter, fuel pipe, and injectors) as well as in predicting air–fuel mixing phenomena and combustion characteristics.

A traditional statistical technique of linear regression (principle of least squares) was used to estimate the flash point, fire point, density and viscosity of diesel and biodiesel mixtures.[23] A set of seven neural network architectures, three training algorithms along with 10 different sets of weights and biases were examined to choose the best Artificial Neural Network (ANN) to predict the properties of diesel–biodiesel mixtures. The performance of both the traditional linear regression and the ANN techniques were then compared to check their validity for predicting the properties of various mixtures of diesel and biodiesel. The results show that the ANN model is a better choice for this particular system. The predicted and experimental values of the properties have a negligibly small error in the case of ANN and it gives a better estimation of said properties than the statistical technique of curve fitting (principle of least squares). It can be

inferred that the neural network NN7 (2-7-4) along with the Levernberg–
Marquardt algorithm can be a better choice over the principle of least
squares to predict the aforementioned properties of various mixtures of diesel
and biodiesel. The performance of neural networks may further be improved
by adjusting the other training parameters like goal, epochs, learning rate,
magnitude of the gradient *etc.*

Krisnangkura *et al.* used the concept of the free energy of viscous flow
(ΔG_{vis}) to predict the viscosity.[24] In this study, ΔG_{vis} for a non-associated
liquid mixture is assumed to be the sum of the ΔG_{vis} values of individual
components. The equation:

$$V = Ae^{-G_{vis}/RT} \text{ is transformed to } \ln V_{blend} = a + bn_1 + c/T + dn_1/T \qquad (22)$$

where *a*, *b*, *c*, *d*, *T* and n_1 are thermodynamically related constants, absolute
temperature and the mole fraction of biodiesel, respectively. The transformed equation is used to predict the kinematic viscosity of biodiesel
blends (V_{blend}) of different degrees of blending at any temperature from the
pour point to 100 °C. The predicted kinematic viscosities are in good
agreement with those reported in the literature at all temperatures. The
highest deviation is $\pm 5.4\%$ and the average absolute deviation (AAD) is
less than 2.86%. The transformed equation can also be used to predict the
kinematic viscosities of pure fatty acid methyl esters in diesel fuel. The
numeric values of *a*, *b*, *c* and *d* (for commercial biodiesel and low-sulfur
petrodiesel blends) are -6.26, 0.459, 2283.7 and -35.96, respectively. Thus
eqn (23) is obtained by substituting these four numeric values into eqn (22)
and the equation can be used to estimate the kinematic viscosities of
biodiesel blends of different degrees of blending at any temperature above
the pour point to 100 °C.

$$\ln V_{blend} = -6.26 + 0.459n_1 + 2283.7/T - 35.96n_1/T \qquad (23)$$

A model predicting the viscosity of diesel–biodiesel mixtures has also been
developed.[25] This was based on existing correlations capable of providing
quality characteristics for the mixtures. The model was also used to maximize
the biodiesel fraction in the diesel– biodiesel mixtures, while taking into
consideration all product quality specifications as they are defined by Greek
legislation. The model was developed in MATLAB and the corresponding
biodiesel optimization studies were carried out with MATLAB's optimization
toolbox.

Ceriani's model predicts the viscosity of fatty acid esters based on a group
contribution method; *i.e.*, a compound or a mixture of compounds is
considered as a solution of groups, and its properties are the sum of the
contributions of each group.[26] The model for the pure compounds is described
in eqns (24)–(26).

$$\ln V = \sum N_k(A_{1k} + B_{1k}/T + C_{1k} \ln T - D_{1k}T)$$
$$- \left[M_i \sum N_k(A_{2k} + B_{2k}/T - C_{2k} \ln T - D_{2k}T) \right] + Q$$

$$\ln V = \sum N_k(A_{1k} + B_{1k}/T + C_{1k} \ln - D_{1k}T)$$
$$- \left[M_i \sum N_k(A_{2k} + B_{2k}/T - C_{2k} \ln T - D_{2k}T) \right] + Q$$

(24)

$$Q = (f_0 + N_C f_1)q + (s_0 + N_{CS} s_1) \tag{25}$$

$$q = \alpha + \beta/T - \gamma \ln(T - \delta T) \tag{26}$$

where V has the units mPa s, T is measured in K, N_k is the number of groups k is the molecule i, M_i is the component molecular weight, A_{1k}, B_{1k}, C_{1k}, D_{1k}, A_{2k}, B_{2k}, C_{2k}, and D_{2k} are parameters obtained from the regression of the experimental data, Q is a correction term, f_0, f_1, s_0, and s_1 are optimized constants, R, β, γ, and δ are optimized parameters obtained by regression of the databank as a whole, N_C is the total number of carbon atoms in the molecule, and N_{CS} is the number of carbons in the alcohol side-chain.

A model that could predict the value of the viscosity of a biodiesel based on the knowledge of its composition would be useful in the optimization of biodiesel production processes and the planning of blending of raw materials and refined products.[27] This work proposed a revised version of Yuan's model proposed here.[17] The results for several biodiesel systems show that the revised Yuan's model provides the best description of the experimental data with an average deviation of 4.65%, as compared to 5.34% for Yuan's original model. The same conclusions were obtained when applying these models to predict the viscosity of blends of biodiesel with petrodiesel.

Ceriani *et al.* proposed a group-contribution method for estimation of the viscosity of fatty compounds and biodiesel esters as a function of the temperature.[28] The databank used for regression of the group-contribution parameters (1070 values for 65 types of substances) included fatty compounds, such as fatty acids, methyl and ethyl esters, alcohols, tri- and di-acylglycerols, and glycerol. The inclusion of new experimental data for fatty esters, a partial acylglycerol, and glycerol allowed for further refinement in the performance of this methodology in comparison to a prior group-contribution equation. In addition, the influence of small concentrations of partial acylglycerols, intermediate compounds in the transesterification reaction, on the viscosity of biodiesels was also investigated. The group-contribution equation chosen to correlate the viscosity V, in mPa s, of component i, and the temperature T in

K, has a different temperature dependency relative to the model proposed earlier:[26]

$$\ln V = \sum N_k(A_{1k} + B_{1k}/T + C_{1k}) + \left[M_i \sum N_k(A_{2k} + B_{2k}/T + C_{2k})\right] + Q \quad (27)$$

where N_k is the number of groups k in the molecule i, M_i is the component molecular weight, A_{1k}, B_{1k}, C_{1k}, A_{2k}, B_{2k} and C_{2k} are parameters obtained from the regression of the experimental databank, k represents the groups of component i, and Q is a correction term.

The specific gravity and viscosity of biodiesel fuels are key properties in determining the suitability for use of such fuels in diesel engines.[29] Correlation models were developed to predict the variation of specific gravity and viscosity with both temperature and % biodiesel in a blend. Blends of jatropha methyl ester (JME) and DF2 (diesel fuel-2) on the basis of volume were prepared and investigated in the temperature range 15–60 °C together with the pure fuels. Blends B20, B35, B50 and B75 were made. Correlation models were proposed to approximate the viscosity of any blend as a function of the temperature in the range from 15 to 60 °C, and as a function of the % of biodiesel.

The variation of viscosity with temperature was accurately fitted using a polynomial model of the form of eqn (28):

$$V = aT^2 - bT + c \quad (28)$$

where V is the viscosity in cSt, T is the temperature in °C and a, b and c are constants.

Another model was in the form of eqn (29):

$$V = a - 0.03T \quad (29)$$

The logarithmic model was in the form of eqn (30):

$$V = -b \ln T + c \quad (30)$$

where V is the viscosity in centistokes, T is temperature in °C and a, b and c are constants. The variation of viscosity with percentage biodiesel in the blend depicts a linear relationship especially for temperatures above 25 °C as shown in the form of eqn (31):

$$V = mB + k \quad (31)$$

where V is the kinematic viscosity, and m and k are correlation constants.

Many properties of biodiesel, the monoalkyl esters of vegetable oils, animal fats or other tri-acylglycerol-containing feedstocks, are largely determined by their major components, the fatty acid alkyl esters.[30] Viscosity, which affects the flow and combustion of a fuel, is such a property. To gain a better understanding of kinematic viscosity, this work additionally reported data on esters with an odd number of carbons in the fatty acid chain and some unsaturated fatty esters. Furthermore, the kinematic viscosity of some biodiesel fuels is affected by components that are solids at 40 °C. A method based on polynomial regression for determining the calculated viscosity contribution (CVC) of esters that are solid at 40 °C (saturated esters in the C20–C24 range) or esters that are liquids but not available in pure form was presented, as these values are essential for predicting the kinematic viscosity of mixtures containing such esters. The kinematic viscosity data of esters were compared to those of aliphatic hydrocarbons in the C6–C18 range and those of dimethyl diesters. The increase of kinematic viscosity with increasing number of CH_2 groups in the chain was non-linear and depended on the terminal functional groups, chain length and number of double bonds. To illustrate this effect, carbon–oxygen equivalents (COE) were used in which the numbers of carbon and oxygen atoms were added together. A straightforward equation, taking into account only the amounts and kinematic viscosity values of the individual neat components, sufficed to predict the viscosity of mixtures of fatty esters (biodiesels) at a given temperature.

Thus the equation:

$$V_{mix} = \sum A_C \cdot V_C \tag{32}$$

in which V_{mix} = the kinematic viscosity of the biodiesel sample (a mixture of fatty acid alkyl esters), A_C = the relative amount (%/100) of the individual neat ester in the mixture (as determined by, for example, gas chromatography (GC)) and V_C = the kinematic viscosity of these esters, can be used to the predict the kinematic viscosity of a mixture such as a biodiesel at a given temperature if the kinematic viscosity values of the individual compounds are known.

In another study, pseudo-binary biodiesel + diesel fuel and biodiesel + benzene mixtures were prepared and the densities, viscosities, and refractive indices of mixtures were measured at 298.15 K.[31] The accuracy of the different mixing rules and empirical equations used to estimate these properties was evaluated. The viscosity of biodiesel + petroleum diesel fuel can be predicted with a very good accuracy using the empirical equations.

Mixtures of biodiesel and ultra-low-sulfur diesel (ULSD) were used to study the variation of density and kinematic viscosity as a function of percent volume (v) and temperature (T), experimental measurements were carried out for six biodiesel blends at nine temperatures in the range 293.15–373.15 K.[32] Viscosity increases because of the increase in the concentration of biodiesel in the blend, and both the viscosity and density decrease as the temperature increases. Empirical correlations were developed to predict the kinematic viscosity, where γ, ϕ and ω are adjustable parameters:

$$V = e^{\ln \gamma + \varphi v + \omega/T + \lambda v/T^2} \tag{33}$$

$$V = e^{\ln \gamma + \omega/T + \lambda v/T^2} \tag{34}$$

$$V = e^{\ln \gamma + \omega/T + \lambda v/T} \tag{35}$$

The estimated values of viscosity were in good agreement with the experimental data; absolute average prediction errors of 0.02 and 2.10% were obtained in the biodiesel(1) + ULSD(2) system studied in this work.

Researchers have also proposed a method to predict the dynamic viscosity values of hypothetical acidic oil-based biodiesel fuels.[33] A topological index (χ_m) that can formulate the molecular structural information of FAME components into a dimension-less form was built from the integration and modification of the distance matrix and adjacency matrix of the molecular structure. A regression equation that correlates the mean topological index values and the dynamic viscosity values was expressed as eqn (36), which allowed the dynamic viscosity values (313 K) of the hypothetical acidified oil-based biodiesel fuels to be calculated from the molecular structures of their FAME components.

$$V_m = 5.96 \ln \chi_m - 10.09 \tag{36}$$

Results show that both caprylic acid (C8 : 0) and linoleic acid (C18 : 2) are beneficial components for reducing biodiesel's dynamic viscosity. Nine hypothetical acidic oil-based biodiesel fuels were made by mixing C8 : 0 or C18 : 2 with one vegetable oil at the same mass ratio, including peanut, canola, coconut, palm, soybean, corn, rapeseed, sunflower and cotton seed oils, and the dynamic viscosity values of each blend were predicted. The dynamic viscosity values were ranged between 0.90–2.13 mPa s (with the addition of C8 : 0) and 2.83–4.07 mPa s (with the addition of C18 : 2).

Recently, a new model for predicting the viscosities of biodiesel has been presented.[34] This model is based on the principle of corresponding states, using one- and two-reference fluids. The best result was found using a two-reference-fluid model, methyl laurate (C12 : 0) and methyl oleate (C18 : 1), with a global average relative deviation of 6.66%.

Empirical equations were developed to estimate four physical properties of methyl esters, and average absolute deviations of 5.95, 2.57, 0.11 and 0.21% for the cetane number, kinematic viscosity, density, and higher heating value were found.[35] Viscosity increases because of the increase in molecular weight and decreases as the number of double bonds increases. The kinematic viscosity expressed as a function of M_i and N is:

$$\ln V_i = -12 - 503 + 2.496 \ln M_i - 0.178N \tag{37}$$

where M_i represents the molecular weight of the ith FAME, and N is the number of double bonds in a given FAME, therefore N is equal to zero, one, two, and three for the methyl esters C18 : 0, C18 : 1, C18 : 2 and C18 : 3, respectively. The prediction of the kinematic viscosity of biodiesel by eqn (37) produces results that are very close to the experimental values.

Viscosity correlates more strongly with the degree of unsaturation, with higher unsaturation leading to lower viscosities.[36] Furthermore, the double bond configuration influences viscosity, with the *trans* configuration giving a higher viscosity than the *cis*.[3,4] Most natural oils are dominated by *cis* double bonds, but some yellow grease (waste cooking oils) can have substantial levels of the *trans* configuration.[4] The location of the double bond within the fatty acid chain has little influence on viscosity.

6.3 Prediction of Sulfur Content

A model predicting sulfur content along with other properties of diesel–biodiesel mixtures was developed. This was based on existing correlations capable of providing quality characteristics for the mixtures.[25] The model was developed in MATLAB and the corresponding biodiesel optimization studies were carried out with MATLAB's optimization toolbox. The property predictions were consistent with the expected quality of the mixtures and with the effect of increasing the biodiesel mixing ratio. The same model was also utilized as the basis for the constrained optimization in maximizing the biodiesel mixing ratio. This optimization methodology was able to identify a maximum biodiesel mixing ratio of 1.97% v/v for which all property specifications are maintained within their limits.

Partial least squares (PLS) models using near- and mid-infrared spectrometry were developed to predict the sulfur content and other quality parameters of diesel–biodiesel blends.[37] Practical aspects were discussed, such as calibration set composition; model efficiency using different infrared regions and spectrometers; and the calibration transfer problem. The root-mean-square errors of prediction, employing both regions and equipment, were comparable with the reproducibility of the corresponding standard method for the properties investigated. Calibration transfer between the two instruments, using direct standardization (DS), yielded prediction errors comparable to those obtained with complete recalibration of the secondary instrument. The results demonstrate that PLS global models employing near- and mid-infrared regions can be used to predict sulfur content in biodiesel–diesel blends.

6.4 Prediction of Flash Point

Flash point is inversely related to fuel volatility. The biofuel specifications for flash point are meant to guard against contamination by highly volatile impurities—mainly excess methanol remaining after product-stripping processes.[36]

In order to improve upon the physicochemical properties of biodiesel, methyl, ethyl, isopropyl and *n*-butyl esters of β-branched fatty acids have been synthesized, initiating from β-branched alcohols.[16] β-Branched alcohols upon oxidation gave corresponding acids, which were converted to their esters. The synthesized esters met all of the required fuel specifications for biodiesels except for flash point. The flash points of methyl, ethyl, isopropyl and butyl esters are 60.8, 64.0, 80.0 and 96.5 °C, respectively. The lower flash point of branched esters is lower than that of biodiesel and can be attributed to the isomerization in the structure. The reduction of flash point in tall oil biodiesel has also been reported, due to the presence of manganese and nickel additives.[38] However, the flash point of branched esters is much higher than the required limits for the flash point of mineral diesel, *i.e.*, 35 °C, and that of aviation fuels. The flash point also increases with the increasing carbon length of the alcohol used for esterification. Therefore, the flash point may be improved further by optimizing the carbon chain length of the branched acids.

The volatility and flash point for pseudo-binary mixtures of sunflower-seed-based biodiesel + ethanol were measured over the entire composition range.[39] The vapor pressures of mixtures of biodiesel + ethanol as a function of temperature were measured by comparative ebulliometry with an inclined ebulliometer. It was found that ethanol can effectively adjust the volatility and flash point of biodiesel. The correlation of the flash point with vapor pressure data for pseudo-binary mixtures of biodiesel + ethanol displays agreement with experimental values.

$$1 = \sum p_i x_i y_i / p_i, \quad FP = p_1 x_1 y_1 / p_1, \quad FP = p_2 x_2 y_2 / p_2 \quad (38)$$

where p_i is the vapor pressure of each pure species, x_i is the liquid mole fraction of a flammable substance i, and y_i is the activity coefficient in the liquid phase.

A traditional statistical technique of linear regression (PLS) was used to estimate the flash point of diesel and biodiesel mixtures.[23] A set of seven neural network architectures, three training algorithms along with 10 different sets of weights and biases were examined to choose best ANN to predict the flash point of diesel–biodiesel mixtures. The performance of both of the traditional linear regression and ANN techniques were then compared to check their validity for predicting the properties of various mixtures of diesel and biodiesel.

The concentrations of key impurities for biodiesel quality and some qualitative parameters like flash point were determined and the relationships between them studied.[21] The relationships are characterized by linear or non-linear statistical models. These models enable a better understanding of the significance of the qualitative parameters and estimate them from the concentrations of impurities. In the case of this research, only residual ethanol was present in biodiesel. Therefore, the flash point was pre-supposed to depend only on the ethanol content.

The correlation equation was:

$$FP = 40.88 - 39.26 \ln(W_{ET} + 0.0194) \quad (R^2 = 0.9927) \tag{39}$$

Where FP is the flash point and W_{ET} is the ethanol content.

At a European symposium, a model predicting the flash point of diesel–biodiesel mixtures was presented.[25] This was based on existing correlations capable of providing quality characteristics for mixtures. The model was developed in MATLAB and the corresponding biodiesel optimization studies were carried out with MATLAB's optimization toolbox.

6.5 Prediction of Cetane Number

Cetane number (CN) is a measure of a fuel's auto-ignition quality characteristics. As biodiesel is largely composed of long-chain hydrocarbon groups (with virtually no branching or aromatic structures) it has a higher CN than petroleum diesel, and increasing the biodiesel level of biodiesel blends increases the CN of the blend.[40,41] Biodiesel produced from feedstocks rich in saturated fatty acids (for example tallow and palm) has a higher CN than fuels produced from less saturated feedstocks (for example soy and rapeseed).

The CNs of 29 samples of straight-chain and branched C1–C4 esters as well as 2-ethylhexyl esters of various common fatty acids were determined.[42] The CNs of these esters are not significantly affected by branching in the alcohol moiety. Therefore, branched esters, which improve the cold-flow properties of biodiesel, can be employed without greatly influencing ignition properties compared to the more common methyl esters. Unsaturation in the fatty acid chain was again the most significant factor causing lower CNs. CNs were determined in an ignition quality tester (IQT) which is a newly developed, automated rapid method using only small amounts of material. The IQT is as applicable to biodiesel and its components as previous cetane-testing methods were. CN was plotted *vs.* the ignition delay time (ID) time for the present compounds and showed a non-linear relationship between the ID time and the CN. The relationship between ID and CN for the IQT is provided by the following equation:

$$CN = 83 : 99(ID - 1 : 512)^{-0.658} + 3 : 547 \tag{40}$$

Thus, small changes at shorter ID times result in greater changes in CN than at longer ID times. While this could indicate a leveling-off effect on emissions once a certain ID time with corresponding CN has been reached, as the formation of certain species depends on the ID time, the relationship between the CN and engine emissions is complicated by many factors including the technology level of the engine.

Researchers have developed a relationship between the FAME composition and the CN.[43] Data were collected from the literature on the CNs of various biodiesel fuels and their FAME composition. A nine-by-eight matrix was formed with CN as the dependent variable and the pure FAME composition as the independent variables. A linear regression analysis was performed on the average values of the CN data of the pure FAME.

The regression equation was as follows:

$$CN = K + ax_1 + bx_2 + cx_3 + dx_4 + ex_5 + fx_6 + gx_7 + hx_8 \qquad (41)$$

where CN = cetane number, K, a, b, c, d, e, f, g and h are constants to be determined by regression analysis, and x_1......x_8 are % compositions of FAMEs. Substituting the values of K and a–h into eqn (41) for the coefficients obtained during regression analysis, eqn (41) becomes:

$$CN = 61.1 + 0.088x_2 + 0.133x_3 + 0.152x_4 - 0.101x_5$$
$$- 0.039x_6 - 0.243x_7 - 0.395x_8 \qquad (42)$$

Eqn (42) shows the relationship between CN and the FAME composition. From the equation, it is observed that the coefficient for the saturated FAMEs is positive and increases with increasing carbon number from C12 : 0 to C16 : 0. This suggests an increase in the CN number with an increase in the composition of saturated FAMEs. However, the coefficient of the unsaturated FAME is negative, therefore a reduction in the overall CN with unsaturation is evident, which further reduces with increasing number of carbons. The same trend is observed for pure methyl esters. This indicates that FAME composition will play a dominant role in establishing the CN. The coefficient of determination (R^2) was 0.88, which indicates that the CN can be predicted with 88% accuracy based on the FAME composition of the biofuel. The predicted CN values compare well with the average measured CN values

In another work, the correlation formulated by another researcher was used, and a good correlation between reported and predicted biodiesel CNs was obtained.[44,45]

$$CN = \sum X_{ME} CN_{ME} \qquad (43)$$

where CN, is the cetane number of the biodiesel, X_{ME} is the weight percentage of each methyl ester and CN_{ME} is the cetane number of each individual methyl ester.

In order to improve upon the physicochemical properties of biodiesel, methyl, ethyl, isopropyl and n-butyl esters of β-branched fatty acid have been synthesized, initiating from β-branched alcohols.[16] The CNs of the synthesized surrogate methyl, ethyl, isopropyl and butyl esters are 62.3, 62.5, 65.8 and 68.8, respectively, and meet all the specifications recommended for diesel and biodiesel. The CNs of branched esters are higher than those for biodiesel, which can be attributed to isomerization of the fatty acid carbon chain. Moreover, the CN increases with the increase in ester alkyl chain length.

Therefore, it was concluded that isomerization and an increase in carbon chain length of ester have a positive impact on CN.

A model capable of predicting the CN along with another 11 properties of diesel–biodiesel mixtures was developed using MATLAB.[25] This was based on existing correlations capable of providing quality characteristics for the mixtures.

Researchers defined a term called the 'Biodiesel Cetane Index', which predicts, with high accuracy, biodiesel CNs:[46]

$$CN = -23.523 + \left(2.366 + 6.299e^{-0.411db}\right)ne^{-0.018n} \quad \text{for methyl esters} \quad (44)$$

$$CN = -23.192 + \left(-3.565 + 14.818e^{-0.185db}\right)ne^{-0.032n} \quad \text{for ethyl esters} \quad (45)$$

where n is the number of carbon atoms in the original fatty acid, and db is the number of double bonds in the acid molecule. Eqn (44) can be used together with eqn (45) for the prediction of CNs. For a given type of alkyl ester (methyl or ethyl), eqns (44) and (45) depend only on the number of double bonds and the acid chain length. From eqn (44), solving for db as a function of the density and the acid chain length and substituting this into (45), an expression for the CN depending on the density and the acid chain length is found, and it is clear that for a given chain length or a given degree of unsaturation, the higher the density the lower the CN.

Recently, empirical equations were developed to estimate the CN of methyl esters and an AAD of 5.95% for the CN was founded.[35] CN increases because of the increase of molecular weight and decreases as the number of double bonds increases. Two general mixing rules and five biodiesel samples were used to study the influence of FAMEs on the properties of biodiesel. The CN of each FAME was obtained from the equation:

$$CN_i = -7 : 8 + 0 : 302\ M_i - 20N \quad (46)$$

where CN_i is the cetane number of the ith FAME, M_i represents the molecular weight of the ith FAME, and N is the number of double bonds in a given FAME, therefore N is equal to zero, one, two, and three for the methyl esters C18 : 0, C18 : 1, C18 : 2 and C18 : 3, respectively. In addition, it has also been proposed though eqn (46) that the CN decreases 20 units for each increase in the number of double bonds of the corresponding FAME. The prediction of the CN is very close to the experimental values.

6.6 Prediction of Carbon Residue

The carbon residue (CR) is one of the most important biodiesel quality criteria.[47] EN 14214 requires modification with a preceding distillation (EN ISO 10370) and determination of the carbon residue with 10% of the distillation residue because of the detection limit decrease and the possibility of direct comparison with fossil diesel fuels. However, this distillation step causes problems.[47,48]

A number of samples of rapeseed oil ethyl ester were prepared by alkaline-catalyzed transesterification under various conditions.[21] The concentrations of the key impurities for biodiesel quality (the concentrations of monoglycerides, diglycerides, triglycerides, free glycerol, ethanol, free fatty acids, water) and some qualitative parameters (flash point, carbon residue, and kinematic viscosity at 40 °C) were determined and the relationships between them studied. The relationships were characterized by linear or non-linear statistical models. Researchers determined the carbon residue of the original biodiesel without the preceding distillation. Thus, the measured values correspond directly with the USA and European biodiesel norms. A linear model of the relationship between CR and potassium ion concentration is:

$$CR = 0.00209K \quad (R^2 = 0.830) \tag{47}$$

where K is the potassium ion content.

Although the R^2 value is smaller than 0.9, the linear dependence is evident at the investigated interval. The dependence is characterized by the absence of a constant term. In the case of this work, the linear regression analysis indicates the insignificant effect of glycerides and free fatty acids on CR. This is probably caused by the investigated concentration intervals—glycerides and free fatty acids significantly influence the CR only at higher concentrations. The absence of the constant term in the linear dependence of CR *vs.* potassium ion content indicates that the CR of purified biodiesel is negligible. Researchers have mentioned additional substances (other then potassium compounds), which can increase the carbon residue.[47,49]

6.7   Prediction of Distillation Temperature

PLS models using near- and mid-infrared spectrometry were developed to predict the distillation temperatures of biodiesel.[37] Practical aspects were discussed, such as calibration set composition; model efficiency using different infrared regions and spectrometers; and the calibration transfer problem. The root-mean-square errors of prediction, employing both regions and equipment, were comparable with the reproducibility of the corresponding standard method for the properties investigated. Calibration transfer between the two instruments, using direct DS, yielded prediction errors comparable to those obtained with complete recalibration of the secondary instrument.

6.8   Prediction of Water Content

Water content was predicted using a model developed in MATLAB and the corresponding biodiesel optimization studies were carried out with MATLAB's optimization toolbox.[25] Available literature for the prediction model of water in biodiesel is scarce.

6.9   Relationship between Free Glycerol and Potassium Contents

The following relation was developed between the free glycerol and potassium contents in biodiesel:[21]

$$W_{FG} = 0.00461K \quad (R^2 = 0.823) \tag{48}$$

where W_{FG} is the free glycerol content and K is the potassium content in biodiesel based on the ethyl ester of rapeseed oil. These models enable a better understanding of the significance of qualitative parameters and estimate them from the concentrations of impurities. Previous work on the linear dependence between the potassium ion and free glycerol contents in the ester phase seems to be valid also for FAMEs.[50–52] This dependence was also confirmed with regard to different separation conditions. The measured data confirm the approximately linear dependence between the potassium ion and free glycerol concentrations, although R^2 is 0.823. However, the value of R^2 indicates that there is a more complex situation which exists in the reaction mixture and probably other unknown parameters have an effect on the free glycerol and potassium ion concentrations. The found dependence is characterized by the absence of a constant term, *i.e.*, it predicts a zero concentration of free glycerol if the concentration of potassium ions is also zero.

6.10   Prediction of Mono-, Di- and Tri-Glyceride Content

A number of samples of rapeseed oil ethyl ester were prepared by alkaline-catalyzed transesterification under various conditions.[21] The concentrations of monoglycerides, diglycerides and triglycerides were determined with other parameters and the relationships between them studied. The relationship between the concentrations of monoglycerides (MGs), diglycerides (DGs) and triglycerides (TGs) in the studied samples of biodiesel were also investigated.

$$W_{MG} = 0.60 + 1.17W_{TG} \quad (R^2 = 0.629) \tag{49}$$

$$W_{DG} = 0.29 + 2.73W_{TG} \quad (R^2 = 0.895) \tag{50}$$

where W_{MG}, W_{DG} and W_{TG} are the mono-, di-, and triglyceride concentrations, respectively in biodiesel based on ethyl esters of rapeseed oil. These relationships depend upon the concentration of glycerides.

6.11 Prediction of Higher Heating Value

Neither the USA nor European biodiesel standards include a specification for heating value. Due to its high oxygen content, it is generally accepted that biodiesel from all sources has about a 10% lower mass energy content (MJ kg^{-1}) than petroleum diesel.[36] It should be emphasized that with several biodiesel types, the data reported for heating values is very sparse. As the fatty acid carbon chain increases in length (for a constant unsaturation level) the mass fraction of oxygen decreases, so the heating value increases.[13] However, this increase in heating value with chain length is not readily apparent.[36]

Mathematical relationships between the higher heating value (HHV) and other properties of various biodiesel fuels have been estimated.[13] The HHV is an important property defining the energy content and thereby efficiency of fuels, such as vegetable oils and biodiesels. HHVs for biodiesel are approximately 41 MJ kg^{-1} which is 10% less than those of petrodiesel fuels. Equations were developed for the calculation of the HHV of vegetable oils and biodiesels from their viscosity (V), density (D) and flash point (FP). The equations between HHV and V are for vegetable oils:

$$HHV = 0.0317V + 38.053 \qquad (51)$$

and for biodiesels:

$$HHV = 0.4625V + 39.450 \qquad (52)$$

with coefficient of regression (R) values of 0.9435, and 0.9677, respectively. The correlations may be used for HHV estimation of mixtures of biodiesels obtained from vegetable oils. The relation between HHV and D is:

$$HHV = -0:0259D + 63.776 \qquad (53)$$

with a coefficient of regression (R) value of 0.7982. There is a considerably low regression between density and higher heating value for biodiesel samples. The correlation between HHV and FP for biodiesels is:

$$HHV = 0.021FP + 32.12 \qquad (54)$$

with coefficient of regression (R) value of 0.9530.

Another model was developed in MATLAB and the HHV was studied using this.[25]

In a recent study, researchers considered HHVs to be one of the important properties which affect the utilization of biodiesel fuels, because they are required as input data for predictive engine combustion models.[35] Empirical equations were developed to estimate the HHV and an AAD of 0.21% for the HHV was found.

The HHV of methyl esters can be calculated from:

$$HHV = 46.19 - 1794/M_i - 0.21N \qquad (55)$$

where *HHV* is the higher heating value of the *i*th FAME in MJ kg^{-1}, M_i represents the molecular weight of the *i*th FAME, and *N* is the number of double bonds in a given FAME, therefore *N* is equal to zero, one, two and three for the methyl esters C18 : 0, C18 : 1, C18 : 2 and C18 : 3, respectively.

6.12   Prediction of Iodine Value

Most common analytical methods in FA chemistry can be categorized as yielding either structure or quality indices.[53] Common structure indices are the iodine value (IV), the saponification value, and the hydroxyl value. Although modern analytical methods yield more detailed and reliable information, structure indices are still widely used. The IV, which indicates total unsaturation, has even been included in standards for biodiesel. However, the IV index is too general to allow the correlation of physical and chemical properties with FA composition. The IV reflects biodiesel- and oxidative-stability-related issues. Alternative indices for the IV are being developed. Possible alternatives are the allylic position equivalent (APE) and the bisallylic position equivalent (BAPE), which better relate the structure and the amount of common component FA in vegetable oils to their observed properties. The APE and BAPE indices are based on the number of reactive positions during the oxidation process.[53]

$$IV = 100(253.81db)/MW_f \qquad (56)$$

$$IV_{mixture} = \sum 100(253.81A_f db)/MW_f \qquad (57)$$

in which *IV* is the iodine value, *db* is the number of double bonds, MW_f is the molecular weight of the fatty compound, 253.81 is the atomic weight of the two iodine atoms that are theoretically added to one double bond, and A_f is the amount (%) of a fatty compound in a mixture.

The IV is directly related to FAME unsaturation.[36] FAME molecules having multiple double bonds. For this reason, there is some controversy about the need for an IV standard at all, and certainly about the rather restrictive maximum IV value of 120 g I_2 per 100 g biodiesel set by EN 14214.

Researchers have also argued that there is no need for an IV specification
because the CN specification effectively limits unsaturation.[54]

Multivariate near-infrared spectroscopy models for predicting the IV has
been developed.[15] Principal component analysis was used to perform a
qualitative analysis of the spectra and a PLS regression was used to develop the
calibration models between analytical and spectral data. The results supports
the notion that near-infrared spectroscopy, in combination with multivariate
calibration, is a promising technique to be applied to biodiesel quality control,
in both laboratory- and industrial-scale samples.

6.13   Prediction of Lubricity

Lubricity refers to the reduction of friction between solid surfaces in relative
motion.[55] Two general mechanisms contribute to overall lubricity: (a)
hydrodynamic lubrication and (b) boundary lubrication. In hydrodynamic
lubrication, a liquid layer (such as diesel fuel within a fuel injector) prevents
contact between opposing surfaces. Boundary lubricants are compounds that
adhere to the metallic surfaces, forming a thin, protective anti-wear layer.
Boundary lubrication becomes important when the hydrodynamic lubricant
has been squeezed out or otherwise removed from between the opposing
surfaces. Good lubricity in diesel fuel is critical to protect fuel-injection
systems. In many cases, the fuel itself is the only lubricant within a fuel
injector. With the increasing operational demands of modern injection
systems—due to higher pressures, injection rate shaping, multiple injections
per cycle, and other features—maintaining adequate lubricity is more critical
than ever. Biodiesel from all feedstocks is generally regarded as having
excellent lubricity, and the lubricity of ULSD can be improved by blending
with biodiesel.[36] Because of its naturally high lubricity, there is no lubricity
specification for B100 within either the USA or European biodiesel standards.
Biodiesel's good lubricity can be attributed to the ester group within the
FAME molecules, but a higher degree of lubricity is due to trace impurities in
the biodiesel. In particular, free fatty acids and monoglycerides are highly
effective lubricants.[56] The effect of unsaturation upon lubricity is unclear, with
some researchers reporting positive effects of C=C bonds while others report
no effect.[56,57] Efforts to reduce impurities like monoglycerides to improve low-
temperature properties could lead to a worsening of the lubricity.

Lubricity is considered to be one of the important fuel properties of biodiesel
that are influenced by the fatty acid profile and, in turn, by the structural
features of the various fatty esters.[58] Studies on the lubricity of biodiesel or
fatty compounds have shown the beneficial effect of these materials on the
lubricity of conventional petroleum-derived diesel fuels, particularly low-sulfur
petrodiesel fuel.[59-65] Adding biodiesel at low levels (1–2%) restores the
lubricity to low-sulfur petroleum-derived diesel fuels.[58] Unsaturated acids
exhibited better lubricity than saturated species and ethyl esters had improved
lubricity compared to methyl esters.[59,66]

6.14   Prediction of Density

Density can be defined as the mass of an object divided by its volume. Fuel density is a key property that affects engine performance.[36] Because fuel-injection pumps meter fuel by volume, not by mass, a greater or lesser mass of fuel is injected depending upon its density. Thus, the air-to-fuel ratio and energy content within the combustion chamber are influenced by fuel density. The densities of biodiesel fuels are slightly higher than those of petroleum diesel, and increasing the biodiesel level of biodiesel blends increases the blend's density.

In the work by Alptekin and Canakci, two different commercially available diesel fuels were blended with the biodiesels produced from six different vegetable oils (sunflower, canola, soybean, cotton seed, corn oils and waste palm oil).[14] The blends (B2, B5, B10, B20, B50 and B75) were prepared on a volume basis. The key fuel properties such as density and viscosities of the blends were measured by following ASTM test methods. Eqn (58) fits the data sufficiently well that higher degree equations are not required:

$$D = Ax + B \qquad (58)$$

where D is density (g cm^{-3}), A and B are coefficients and x is the biodiesel fraction. The calculated density values from eqn (58) were validated by using the measured density values for all the blends. There is strong agreement between the measured and estimated values. The maximum absolute error between the measured and estimated values is 0.42% and the minimum R^2 is 0.9984 for biodiesel–N.D fuel blends. The maximum difference between the measured values and estimated values and the minimum R^2 are 0.20% and 0.9996 for biodiesel–S.E fuel blends, respectively.

The model developed in MATLAB was also used to predict the densities of biodiesels.[25]

A traditional statistical technique of linear regression (PLS) was used to estimate the density of diesel and biodiesel mixtures.[23] A set of seven neural network architectures, three training algorithms along with 10 different sets of weights and biases were examined to choose the ANN to best predict the density. The performances of both of the traditional linear regression and ANN techniques were then compared to check their validity to predict the properties of various mixtures of diesel and biodiesel.

In another study, near-infrared spectroscopy was used to determine the density at 15 °C.[15] Principal component analysis was used to perform a qualitative analysis of the spectra, and PLS regression was used to develop the calibration models between analytical and spectral data. The results support the notion that near-infrared spectroscopy, in combination with multivariate calibration, is a promising technique to be applied to biodiesel quality control, in both laboratory- and industrial-scale samples. An accurate knowledge of biodiesel density permits the estimation of other properties such as the CN, whose direct measurement is complex and presents low repeatability and low reproducibility.[46] In this study densities of methyl and ethyl esters published in the literature, were complied and equations were proposed to convert them to 15 °C and to predict the biodiesel density based on its chain length and unsaturation degree. Calculations also proved that the introduction of high-biodiesel-content blends in the fuel market would force the refineries to reduce the density of their fossil fuels. Most of the density data obtained from the literature were measured at 20, 25 and 40 °C. Three options have been studied for converting all data to 15 °C (because this is the temperature specified in the biodiesel standards). In order to consider the effect of temperature on the density of alkyl esters, the following equation was proposed:[46]

$$D = D(T)(0.29056 - 0.08775\omega)\left[(1 - T/T_c)^{2/7} - (1 - 288.15/T_c)^{2/7}\right] \quad (59)$$

where D is determined at 288.15 K. $D(T)$ and T_c is temperature and T_c is critical temperature in °C and ω is acentric factor. However, this equation was originally proposed for a saturated liquid over a wide temperature range (up to the critical temperature); therefore its application to a liquid at a lower temperature than its saturation temperature (at a given pressure) is not correct. Moreover, in the range 0–50 °C the equation underestimates the decrease of the density with temperature compared to that derived from experimental density values of different methyl esters. Second, a linear correlation was proposed specifically for biodiesel fuels in EN 14214 in 2009. This is given as:

$$D = D(T) + 0.723(T - 15) \quad (60)$$

where D is determined at 15 °C. However, this is being currently revised by the European Committee for Standardization (Working Group 24) since it overestimates the decrease with temperature and does not account for the different nature of the esters like the type of alkyl ester, the number of double bonds and the chain length. Third, considering the shortcomings of the aforementioned methods, a new equation [eqn (61)] has been developed.[46] Similar to eqn (60) a linear dependence of the density and temperature is considered.

Therefore, the proposed equation is linear:

$$D = D(T) + a(T - 15) \quad (61)$$

where D is determined at 15 °C. The coefficient a is dependent on the type of ester. The final correlation proposed here is a five-coefficient function which depends on the number of carbon atoms in the original fatty acid (n), the number of carbon atoms in the original alcohol used for the transesterification process (m) and the number of double bonds in the acid molecule (db):

$$D = 851.471 + [250.718db + 280.899 - 921.180(m - 1)]/1.214 + n \quad (62)$$

The fit of this equation to the collected data provides a correlation coefficient $R^2 = 0.969$.

In similar work, the aim was to present new density data for different biodiesels and use the reported data to evaluate the predictive capability of models previously proposed to predict biodiesel or FAME densities.[67] Densities were measured for 10 biodiesel samples, for which the detailed composition was reported, at atmospheric pressure and temperatures from 278.15 to 373.15 K. The density dependence with temperature was proposed for the biodiesels, and isobaric expansivities were presented. It was shown that Kay's mixing rules and a revised form of the group-contribution volume (GCVOL) model are able to predict biodiesel densities with average deviations of only 0.3%.[67,68] It was shown that it can predict the densities of biodiesel fuels with average deviations less than 0.4%. Kay's mixing rules are the simplest form of mixing rules by which mixture properties are obtained by summing the products of the component properties using weighting factors, which are usually the concentrations of the components in a mixture. For example:

$$D = \sum m_i c_i p_i \tag{63}$$

where c_i is the concentration of component i. The major drawback in the application of linear mixing rules is that they require knowledge of the experimental densities of the pure components present in the mixture and assume that the mixture excess volumes are negligible. This may not be easy for many real fluids because they are either composed of a large number of compounds or have different natures, and subsequently, the excess volumes are non-negligible. However, biodiesels are simple mixtures composed, in general, of fewer than 10 fatty acid esters all from the same family, and thus, excess volumes are very small. The density data measured here were correlated using a linear temperature dependency and an optimization algorithm based on the least-squares method:

$$D = bT + a \tag{64}$$

where T is measured in K, and the parameter values along with their confidence limits were estimated.

In another study, to predict the density, a mixing rule was evaluated as a function of the volume fraction of biodiesel in the blend.[69] The effects of the biodiesel fraction on each of these properties in addition to the effects of temperature on the density and viscosity were investigated. The blends (B2, B5, B10, B20, B50 and B75) were prepared on a volume basis. Generalized equations and the Arrhenius equation for predicting the density of the blends were used. The low values of the AADs and the maximum absolute deviations obtained confirmed the suitability of the mixing rule used. For all the blends, it was observed that the results from the measured and estimated values of density and viscosity were in good agreement. From the results, the density of the blends decreased with increasing temperature while these properties increased with increasing fraction of biodiesel in the fuel blend. The measured density of the biodiesels and their blends with commercial grade No. 2 diesel was correlated as a function of biodiesel fraction and temperature, respectively using the linear square method. The linear regression equation formulated is as follows:

$$D = a(T + b) \tag{65}$$

where T is the temperature (°C), D is the density (kg m^{-3}) and a and b are correlation coefficients.

Biodiesel–diesel fuel mixtures have been commonly used in recent years especially in the transport industry, in order to reduce environment pollution, and dependency on imported fossil fuels.[31] Some of the basic properties of these mixtures, especially density and viscosity, strongly influence spray properties, atomization and combustion processes, engine deposits formation, engine behavior in cold weather conditions, and are used as input data for predictive engine combustion models. In this study, pseudo-binary biodiesel–diesel fuel and biodiesel–benzene mixtures were prepared and the densities, viscosities, and refractive indices of mixtures were measured at 298.15 K. The accuracy of the different mixing rules and empirical equations used to estimate these properties was evaluated. The density of the studied mixtures could be predicted with very good accuracy using Kay's mixing rule or empirical equations obtained from regression analysis. Density measurements were carried out at atmospheric pressure and 298.15 K according to the ASTM D-4052 test method and using an Anton Paar DM4500 density meter.

The Kay equation is:

$$D_m = v_1 D_1 + v_2 D_2 \tag{66}$$

where D_m is the density of the mixture (g cm^{-3}), D_1 and D_2 are the densities of components 1 and 2 of the mixture (g cm^{-3}), and v_1 and v_2 are the volume fractions of components 1 and 2. An increased accuracy of density correlation can be obtained using empirical polynomial equations obtained from regression analysis of the measured values:

$$D_m = a v_1 + b \tag{67}$$

$$D_m = a v^2 + b v_1 + c \tag{68}$$

where a, b and c are the regression coefficients.

Using eqn (67), the densities for mixtures of seven types of biodiesels with diesel fuel were predicted, and errors were in the range 0–2%. The experimental data for pseudo-binary mixtures of rapeseed oil biodiesel–diesel fuel were correlated as a function of biodiesel volume fraction in order to determine the values of regression coefficients for the first-order equation, eqn (67). Regression analysis was performed to determine the equation which best fits the experimental data in the case of rapeseed oil biodiesel–benzene mixtures. For this system, eqn (68) was used because it gave better results than eqn (67). The experimental data were correlated with the empirical eqns (67) and (68). The fitting coefficients from eqn (67) were $a = 0.0388$ and $b = 0.8376$ ($R^2 = 0.9997$), and the fitting coefficients from eqn (68) used to calculate the density of biodiesel–benzene mixtures were $a = 0.0024$, $b = 0.0008$ and $c = 0.8736$ ($R^2 = 0.9971$). The correlation between density and composition given by the empirical eqns (67) and (68) is very high. The calculated density of the two pseudo-binary mixtures from the densities of the mixture components, and the mixture compositions had a good accuracy: 0.012 and 0.014% for biodiesel–diesel fuel mixtures and 0.022 and 0.003% for biodiesel–benzene mixtures in terms of AAD. It may be noted that for the biodiesel–diesel fuel system, only two parameters were sufficient, while for the biodiesel–benzene system, three parameters were needed for density correlation.

Another empirical correlation was proposed to estimate the density:[32]

$$D = aV + bT + d \qquad (69)$$

where T is the temperature in K, and b and d are adjustable parameters.

The estimated values of the density were in good agreement with the experimental data and absolute average prediction errors of 0.02 and 2.10% were obtained for the biodiesel(1) + ULSD(2) system studied in this work.

PLS models using near- and mid-infrared spectrometry were developed to predict the density of diesel–biodiesel blends.[37] Practical aspects were discussed, such as calibration set composition; model efficiency using different infrared regions and spectrometers; and the calibration transfer problem. The root-mean-square errors of prediction, employing both regions and equipment, were comparable with the reproducibility of the corresponding standard method for the properties investigated. Calibration transfer between the two instruments, using DS, yielded prediction errors comparable to those obtained with complete recalibration of the secondary instrument.

Density is considered to one of the important properties that affects the utilization of biodiesel fuels, because it is involved in the definition of fuel quality and required as input data for predictive engine combustion models.[35] In this work, the researchers presents the characterization of two biodiesel samples made from beef tallow and soybean oil through their FAME profile. Empirical equations were developed to estimate the density of the methyl esters; and an AAD of 0.11% density was obtained. The density decreases as the molecular weight increases and density increases as the degree of unsaturation increases. Two general mixing rules and five biodiesel samples were used to study the influence of FAMEs over the physical properties of biodiesel. The prediction of the density of biodiesels was very close to the experimental values. The expression for the density of saturated and unsaturated FAMEs is:

$$D_i = 0.8463 + 4.9/M_i + 0.018N \tag{70}$$

where D_i is the density at 20 °C of the ith FAME in g cm^{-3}, M_i represents the molecular weight of the ith FAME, and N is the number of double bonds in a given FAME, therefore N is equal to zero, one, two and three for the methyl esters C18 : 0, C18 : 1, C18 : 2 and C18 : 3, respectively.

References

1. G. Knothe and K. R. Steidley, *Fuel*, 2007, **86**, 2560.
2. B. Kegl, *Fuel*, 2008, **87**, 1306.
3. G. Knothe, in *The Biodiesel Handbook*, ed. G. Knothe, J. V. Gerpen and J. Krahl, AOCS Press, Urbana, 2005, p. 81.
4. A. A. Refaat, *Int. J. Environ. Sci. Technol.*, 2009, **6**, 677.
5. L. Grunberg and A. H. Nissan, *Nature*, 1949, **164**, 799.
6. W. D. Monnery, W. Y. Svrcek and A. K. Mehrotra, *Can. J. Chem. Eng.*, 1995, **73**, 3.
7. J. B. Irving, Report No. 630National Engineering Laboratory, London, 1977.
8. C. A. W. Allen, K. C. Watts, R. G. Ackman and M. J. Pegg, *Fuel*, 1999, **78**, 1319.
9. G. Knothe and K. R. Steidley, *Fuel*, 2005, **84**, 1059.
10. R. M. Joshi and M. J. Pegg, *Fuel*, 2007, **86**, 143.
11. Q. Shu, B. Yang, J. Yang and S. Qing, *Fuel*, 2007, **86**, 1849.
12. G. Knothe, *Energy Fuels*, 2008, **22**, 1358.
13. A. Demirbas, *Fuel*, 2008, **87**, 1743.
14. E. Alptekin and M. Canakci, *Renewable Energy*, 2008, **33**, 2623.
15. P. Baptista, P. Felizardo, J. C. Menezes and M. J. N. Correia, *Talanta*, 2008, **77**, 144.
16. R. Sarin, R. Kumar, B. Srivastav, S. K. Puri, D. K. Tuli, R. K. Malhotra and A. Kumar, *Bioresour. Technol.*, 2009, **100**, 3022.
17. W. Yuan, A. C. Hansen and Q. Zhang, *Fuel*, 2009, **88**, 1120.
18. H. Vogel, *Phys. Z.*, 1921, **22**, 645.
19. R. H. Perry and D. W. Green, *Perry's Chemical Engineers' Handbook*, *7th ed.* Prepared by a staff of specialists under the *editorial* direction of late *editor* Robert H. Perry; *editor*, Don W. Green; assistant *editor*, James O. Maloney. New York, USA, McGraw-Hill, 1997.
20. C. W. Bonhorst, P. M. Althouse and H. O. Triebold, *Ind. Eng. Chem.*, 1948, **40**, 2379.
21. M. Cernoch, M. Hek and F. Skopal, *Bioresour. Technol.*, 2010, **101**, 7397.
22. B. Tesfa, R. Mishra, F. Gua and N. Powles, *Renewable Energy*, 2010, **35**, 2752.

23. J. Kumar and A. Bansal, *Kathmandu Univ. J. Sci. Eng. Technol.*, 2010, **6**, 98.
24. K. Krisnangkura, C. Sansa-ard, K. Aryusuk, S. Lilitchan and K. Kittiratanapiboon, *Fuel*, 2010, **89**, 2775.
25. K. Kalogeras, S. Bezergianni, V. Kazantzi and P. A. Pilavachi, presented at the 20th European Symposium on Computer Aided Process Engineering, Ischia, Italy, 2010.
26. R. Ceriani, C. B. Goncalves, J. Rabelo, M. Caruso, A. C. C. Cunha, F. W. Cavaleri, E. A. C. Batista and A. J. A. Meirelles, *J. Chem. Eng.*, 2007, **52**, 965.
27. S. V. D. Freitas, M. J. Pratas, R. C. A. S. Lima and J. A. P. Coutinho, *Energy Fuels*, 2011, **25**, 352.
28. R. Ceriani, C. B. Gonçalves and J. A. P. Coutinho, *Energy Fuels*, 2011, **25**, 3712.
29. R. K. Kimilu, J. A. Nyang'aya and J. M. Onyari, *ARPN J. Eng. Applied Sciences*, 2011, **6**, 1819.
30. G. Knothe and K. R. Steidley, *Fuel*, 2011, **90**, 3217.
31. N. S. Geacai and O. Iulian, *Renewable Energy*, 2011, **36**, 3417.
32. L. F. Ramírez-Verduzco, B. E. García-Flores, J. E. Rodríguez-Rodríguez and A. del R. Jaramillo-Jacob, *Fuel*, 2011, **90**, 1751.
33. Q. Shu, C. Yu, D. Cai and D. Xiong, *Renewable Energy*, 2012, **41**, 152.
34. F. R. do Carmo, P. M. Sousa Jr, R. S. Santiago-Aguiar and H. B. de Sant'Ana, *Fuel*, 2012, **92**, 250.
35. L. F. Ramírez-Verduzco, J. E. Rodríguez-Rodríguez and A. del R. Jaramillo-Jacob, *Fuel*, 2012, **91**, 102.
36. S. K. Hoekman, A. Broch, C. Robbins, E. Ceniceros and M. Natarajan, *Renewable Sustainable Energy Rev.*, 2012, **16**, 143.
37. L. de Fátima Bezerra de Lira, F. V. C. de Vasconcelos, C. F. Pereira, A. P. S. Paim, L. Stragevitch and M. F. Pimentel, *Fuel*, 2010, **89**, 405.
38. A. Keskin, M. Guru and D. Altiparmak, *Fuel*, 2007, **86**, 1139.
39. Y. Guo, H. Wei, F. Yang, D. Li, W. Fang and R. Lin, *J. Hazard. Mater.*, 2009, **167**, 625.
40. K. J. Harrington, *Biomass*, 1986, **9**, 1.
41. G. Knothe, in *The Biodiesel Handbook*, ed. G. Knothe, J. V. Gerpen and J. Krahl, AOCS Press, Urbana, 2005, p. 76.
42. G. Knothe, A. C. Matheaus and T. W. Ryan III, *Fuel*, 2003, **82**, 971.
43. A. I. Bamgboye and A. C. Hansen, *Int. Agrophys.*, 2008, **22**, 21.
44. M. J. Ramos, C. M. Fernández, A. Casas, L. Rodríguez and Á. Pérez, *Bioresour. Technol.*, 2009, **100**, 261.
45. L. D. Clements, presented at the 3rd Liquid Fuel Conference, Nashville, 1996.
46. M. Lapuerta, J. Rodruez-Ferndez and O. Armas, *Chem. Phys. Lipids*, 2010, **163**, 720.
47. M. Mittelbach and C. Remschmidt, in *Biodiesel, The Comprehensive Handbook*, ed. M. Mittelbach, 1st edn, Martin Mittelbach, Graz, Austria, 2004.

48. K. Komers and J. Machek, presented at the International Conference on Standards and Analysis of Biodiesel, Vienna, 1995.
49. M. Mittelbach and H. Enzelsberger, *J. Am. Oil Chem. Soc.*, 1999, **76**, 545.
50. M. Cernoch, F. Skopal and M. Hek, *Eur. J. Lipid Sci. Technol.*, 2009, **111**, 663.
51. M. Hek and F. Skopal, *Eur. J. Lipid Sci. Technol.*, 2009, **111**, 499.
52. J. Kwiecien, M. Hek and F. Skopal, *Bioresour. Technol.*, 2009, **100**, 5555.
53. G. Knothe, *J. Am. Oil Chem. Soc.*, 2002, **79**, 847.
54. M. Lapuerta, J. Rodriguez-Fernandez and E. F. de Mora, *Energy Policy*, 2009, **37**, 4337.
55. J. Bacha, J. Freel, L. Gibbs, G. Hemighaus, K. Hoekman, J. Horn, J. A. Gibbs, M. Ingham, L. Jossens, D. Kohler, D. Lesnini, J. McGeehan, M. Nikanjam, E. Olsen, R. Organ, B. Scott, M. Sztenderowicz, A. Tiedemann, C. Walker, J. Lind, J. Jones, D. Scott and J. Mills, Chevron Global Marketing, Chevron, San Ramon, CA 94583, USA, 2007. http://www.chevronwithtechron.com/products/documents/Diesel_Fuel_Tech_Review.pdf
56. G. Knothe, SAE Paper No. 2005-01-3672, SAE International, Warrendale, 2005.
57. A. K. Bhatnagar, S. Kaul, V. K. Chhibber and A. K. Gupta, *Energy Fuels*, 2006, **20**, 1341.
58. G. Knothe, *Fuel Process Technol.*, 2005, **86**, 1059.
59. J. A. Waynick, presented at the 6th International Conference on Stability and Handling of Liquid Fuels, Vancouver, B.C., Canada, 1997.
60. H. V. Gerpen, S. Soylu and M. E. Tat, presented at the ASAE-CSAE-SCGR Annual International Meeting, Toronto, Canada, 1999, Paper No. 996134.
61. L. G. Schumacher and J. V. Gerpen, presented at the ASAE International Meeting, Milwaukee, WI, USA, 2000, Presentation No. 006010.
62. G. Anastopoulos, E. Lois, A. Serdari, F. Zanikos, S. Stornas and S. Kalligeros, *Energy Fuels*, 2001, **15**, 106.
63. D. C. Drown, K. Harper and E. Frame, *J. Am. Oil Chem. Soc.*, 2001, **78**, 579.
64. G. Knothe and K. R. Steidley, *Energy Fuels*, 2005, 19, 1192.
65. A. K. Agarwal, J. Bijwe and L. M. Das, *Trans. ASME*, 2003, **125**, 604.
66. E. Kenesey and A. Ecker, *Tribol. Schmier. Tech.*, 2003, **50**, 21.
67. M. J. Pratas, S. V. D. Freitas, M. B. Oliveira, S. C. Monteiro, A. S. Lima and J. A. P. Coutinho, *Energy Fuels*, 2011, **25**, 2333.
68. W. Kay, *Ind. Eng. Chem.*, 1936, **28**, 1014.
69. C. C. Enweremadu, H. L. Rutto and J. T. Oladeji, *Int. J. Phys. Sci.*, 2011, **6**, 758.

CHAPTER 7

Diesel Engine Efficiency and Emissions using Biodiesel and its Blends

7.1 Introduction

The use of vegetable oil as a fuel for compression ignition engines is not a new idea. Rudolph Diesel used peanut oil to fuel diesel engines during the late 1800s. Petroleum-based diesel fuel has been the fuel of choice for the diesel engine for many years due to an abundant supply and low fuel prices. Biodiesel is being evaluated for use as a fuel for diesel engines due to its cleaner burning tendencies, environmental benefits, and for energy security reasons. Diesel engine efficiency is characterized by its performance, combustion and emission parameters. All these parameters will be discussed in detail in this chapter.

7.2 Influence of Biodiesel with Other Factors on Diesel Engine Performance and Combustion Characteristics

Pure biodiesel and its blends can affect diesel engine performance and combustion. The prominent factors are: brake-specific fuel consumption (BSFC), brake-specific energy consumption (BSEC), brake thermal efficiency (BTE), brake power, indicated mean effective pressure (MEP), mechanical efficiency (ME), ignition delay (ID), maximum heat transfer (MHT), and peak cylinder pressure (PCP). These parameters with others will be discussed.

Biodiesel: Production and Properties
Amit Sarin
© Amit Sarin 2012
Published by the Royal Society of Chemistry, www.rsc.org

Different factors can influence the performance and combustion characteristics of diesel engines fueled with biodiesel and its blends.

7.2.1 Contents and Properties of Biodiesel

The properties of pure biodiesel and its percentage content in diesel blends can influence performance and combustion characteristics. Researchers have evaluated the performance of a Compression Ignition (CI) diesel engine with polanga-based biodiesel.[1] One of the parameters evaluated was BTE which is defined as the ratio of the power output to the power introduced during fuel injection; the latter is the product of the injected mass flow rate and the lower heating value. 100% biodiesel performed best, improving the BTE of the engine by 0.1%. A similar trend was shown for the BSEC.

Engine performance tests have shown that mahua oil methyl ester (MOME) as a fuel does not differ greatly from diesel.[2] A slight power loss, combined with an increase in fuel consumption, was experienced. This may be due to the lower heating value of the ester.

Researchers found that the respective average decreases of torque and power of waste frying oil methyl ester (WFOME) were 4.3 and 4.5% due to the higher viscosity and density and lower heating value (8.8%) of WFOME.[3] It was also observed in another study that the brake torque loss was 9.1% for B100 biodiesel relative to D2 diesel at 1900 rpm as a result of the variation in heating value (13.3%), density and viscosity.[4]

The authors found that the torque and power reduced by 3–6% for pure cotton seed biodiesel compared to diesel, and they claimed that the heating value of biodiesel was less 5% than that of diesel.[5] But they found the reason to be difficulties in fuel atomization instead of the loss of heating value.

Usta reported that the blend, D82.5/tobacco seed oil methyl ester (TSOME17.5) provided the maximum increase in torque, power and thermal efficiency.[6] The maximum increase in power occurred at 2000 rpm as 8.74 kW. This value is 3.13% higher than the power 27.84 kW obtained with diesel fuel. Furthermore, the peak thermal efficiency was observed at 2500 rpm as 29.8%. This is approximately 2.02% higher than the thermal efficiency of diesel fuel.

The BTE with biodiesel was about 31.67% for jatropha methyl esters (JME) (100%) where as it was about 31.59% with diesel at 600 kPa bmep (brake mean effective pressure) and similar trends were observed for 20, 40, 60 and 80% biodiesel.[7] The BSEC is slightly higher compared to diesel at all loads and this may be due to the lower calorific value of biodiesel. The peak pressure rate of cylinder pressure rise was similar for biodiesel and its blends as compared to diesel. There was a difference of 0.2 MPa between the peak pressure with JME100 and diesel at full load. This difference decreases for lower blends of biodiesel. Exhaust gas temperature was observed for all blends of JME with diesel. JME100 has shown 404 °C, and diesel 400 °C at full load. BSFC (which is used to designate quantities that have been normalized by dividing by the engine's power, therefore equal to the fuel flow rate divided by the engine power) was higher for all blends of JME due to the lower calorific value of biodiesel. There is a report of reduced BTE and increased SFC, higher exhaust gas temperature, and longer delay at all loads in comparison to diesel operation.[8] During JME operation, higher density and lower heating value fuel was the reason for the increased BSFC and lower thermal efficiency.

Researchers have reported that during a case study on rapeseed ethyl ester (REE), the dynamometer test at 2500 rpm showed 1.8% less power, 8.9% less fuel economy, and 31.8% less opacity when operated on biodiesel in comparison to diesel fuel.[9] Vehicle performance was extremely good and no problems were encountered.

It has been reported that the torque produced for B20 and B40 blends of karanja methyl ester (KME) were 0.1–1.3% higher than that of diesel due to complete combustion of fuel.[10] In the case of B60–B100, it was reduced by 4–23% when compared with diesel fuel for a single-cylinder, four-stroke, direct-injection, water-cooled engine producing 7.5 kW power at 3000 rpm. The BSFC for B20 and B40 was 0.8–7.4% lower than diesel. In the case of B60–B100 the BSFC was 11–48% higher than diesel because of a decrease in the calorific value of the fuel with an increase in the biodiesel percentage in the blends. The BTE was 26.79 and 29.19% for B20 and B40, respectively, which was higher than that of diesel (24.64%). The maximum BTE obtained from B60, B80 and B100 was 24.26, 23.96, and 22.71%, respectively. This is due to the reduction in calorific value and increase in fuel consumption compared to B20.

Kadiyala *et al.* found that the peak cylinder pressure and rate of pressure rise with the use of pungam methyl ester (PuME) and diesel fuel varied only marginally.[11] The heat-release-rate curves and diffused combustion, in the case of PuME at full load running of the engine, were comparatively better than those of diesel fuel.

Experiments have been carried out to estimate the performance and combustion characteristics of a single-cylinder; four-stroke variable-compression ratio multi-fuel engine fuelled with waste cooking oil methyl ester (WCOME) and its blends with standard diesel.[12] Tests were conducted using the fuel blends of 20, 40, 60 and 80% biodiesel with standard diesel, and an engine speed of 1500 rpm, a fixed compression ratio of 21 and at different loading conditions.[12] The performance parameters elucidated included BTE, BSFC, brake power, indicated mean effective pressure, ME and exhaust gas temperature (EGT). The results of the experiment have been compared and analyzed with standard diesel and they confirm a considerable improvement in the performance parameters.

Researchers carried out an experimental study to use raw algae oil and its methyl esters in a Ricardo E6 variable compression ratio engine.[13] The effects on engine speed, engine load output, injection timing of the algae biofuel and engine compression ratio on the engine output torque, combustion noise (maximum pressure rise rate), maximum pressure and maximum heat release were studied. However, its use reduced the engine output torque slightly and increased the combustion noise.

Recently, WFOME was tested and the performance parameters for different WFOME blends were found to be very close to those for diesel.[14] At rated output, the BTE of B50 (50% biodiesel + 50% mineral diesel) was found to be 6.5% lower than that of diesel. For B50, the BSFC observed was 6.89% higher than that of diesel.

It was found that the increase of biodiesel percentage in the blends resulted in a slight decrease of both power and torque over the entire speed range for different blends (B20, B30, B50, B70, B80 and B100) of biodiesel and diesel on a six-cylinder direct-injection (DI) diesel engine.[15] It was reported that the torque decreased with the increase in cotton seed oil methyl ester (CSOME) in the blends (B5, B20, B50, B75 and B100) due to the higher viscosity and lower heating value of CSOME.[16]

The higher viscosity of biodiesel, which enhances fuel spray penetration, and thus improves air–fuel mixing, has been used to explain the recovery in torque and power for biodiesel relative to diesel in some of the literature.[17,18] However, a few authors concluded that the higher viscosity results in power losses, because the higher viscosity decreases combustion efficiency due to poor fuel injection atomization.[3,16]

7.2.2 Engine Load and Speed

It has been observed that the engine power and torque are increased by the application of a low heat rejection (LHR) engine, mainly due to the increased exhaust gas temperatures before the turbine inlet in the LHR engine.[19]

With increase in load, the BSFC of biodiesel decreases.[20–25] One possible reason for this trend could be the higher percentage increase of brake power with load as compared to fuel consumption. However, Gumus and Kasifoglu showed that the BSEC initially decreased with increasing engine load until it reached a minimum value and then increased slightly with further increasing engine load for all kinds of fuels (B5, B20, B50, B100 and diesel).[26]

Researchers have that reported that the BTE was similar to that of diesel at part and full load and that it was better at 400 kPa bmep as compared to diesel for all blends of JME with diesel.[7]

It was reported that the increase in BSFC values at full load was higher than those at partial loads for biodiesel compared to diesel.[27,28] Three thumba oil biodiesel blends (B10, B20 and B30) and diesel were tested on a four-cylinder, DI and Water-cooled (WC) diesel engine covering a wide range of engine speeds.[29] The BSFC initially decreased sharply with the increase in speed up to 2000 rpm, and then remained approximately constant between 2000 and 4000 rpm. For speeds >4000 rpm, the BSFC increased sharply with speed. However, it has also been shown that the BSFC increases with increasing engine speed.[30,31]

7.2.3   Injection Timing and Pressure

Researchers have studied the effects of the engine design parameters *viz.* compression ratio and fuel injection pressure on the performance with regard to parameters such as fuel consumption and BTE with JME as the fuel.[32] It was found that the combined increase of compression ratio and injection pressure increases the BTE and reduces the BSFC.

A high injection pressure produces smaller droplets with lower momentum and hence shallow penetration. Similarly, a low injection pressure results in a larger droplet size and high peak pressure throughout the range of operation of the engine.[33]

Banapurmath *et al.* compared the effect of three injection timings (19, 23 and 27 °CA (Crank angle in degrees)) and the injection of pressure (IOP) on the BTE for honge oil methyl ester (HOME).[34] They found that there was an improvement in the BTE for biodiesel when the injection timing was retarded, and that the highest BTE occurred at 260 bar among all the IOPs tested because the atomization, spray characteristics, and mixing with air were better at a higher injection pressure, which resulted in improved combustion.

It was also concluded that the difference of BTE and BSFC between biodiesel and pure diesel tended to increase with increasing fuel injection pressure.[35] It was observed that the power and torque also increased but the fuel consumption reduced to almost pure diesel levels by reducing the injection advance because it was possible to optimize combustion.[15]

Researchers retarded the injection timing by 3 °CA on a single-cylinder, naturally aspirated (NA), air-cooled (AC), DI diesel engine equipped with a pump–line–nozzle type fuel-injection system, and they observed that the BSFC increased for both B50 and pure RME (rapeseed methyl ester), although the increase was not significant.[36]

7.2.4   Additives

Keskin *et al.* investigated the influence of Mg- and Mo-based fuel additives on diesel engine performance for an engine running on tall oil biodiesel.[37] The authors found that the engine performance values did not change significantly with biodiesel fuels.

Chicken fat biodiesel containing a synthetic Mg additive was studied in a single-cylinder, DI diesel engine and its effects on engine performance and exhaust emissions were studied.[38] A synthetic, organic-based Mg additive was doped into the biodiesel blend at a concentration of 12 μmol Mg. Engine tests were run with diesel fuel (EN 590) and a blend of 10% chicken fat biodiesel and B10 at full-load operating conditions and different engine speeds from 1800 to 3000 rpm. The results showed that, the engine torque was not changed significantly with the addition of 10% chicken fat biodiesel, while the BSFC increased by 5.2% due to the lower heating value of biodiesel.

It was also proved that diethyl ether could be used as an additive in biodiesel to improve its performance characteristics.[39] An experimental investigation was conducted to determine the effect of oxygen content and oxygenate type on DI diesel engine performance and emissions.[40] One conventional and three oxygenated fuels were examined, having oxygen content ranging from 0 to 9%. The fuels were prepared by blending biodiesel, diglyme and butyl-diglyme with a low-sulfur diesel fuel in various proportions. A slight increase of the BSFC was observed due to the small decrease of fuel heating value with the increasing oxygen content. Similar effects were observed when replacing RME with a mixture of diglyme and butyl-diglyme and leaving the oxygen percentage unaltered.

Keskin *et al.* studied the effect of metallic fuel additives on the performance characteristics of diesel engines fuelled with tall oil methyl ester.[41] Metallic fuel additives improved the properties of biodiesel fuels, such as the pour point and viscosity. Biodiesel fuels were tested in an unmodified DI diesel engine at full-load conditions. The BSFC of biodiesel fuels increased by 6%, however, in comparison with B60, it showed a decreasing trend with the use of additives.

Kanna *et al.* investigated the use of ferric chloride ($FeCl_3$) as a fuel-borne catalyst (FBC) for waste cooking palm oil based biodiesel.[42] The metal-based additive was added to biodiesel at a dosage of 20 μmol L^{-1}. Experiments were conducted to study the effect of ferric chloride added to biodiesel on the performance of a DI diesel engine operated at a constant speed of 1500 rpm under different operating conditions. The results showed that the FBC-doped biodiesel resulted in a decrease of the BSFC of 8.6% while the BTE increased by 6.3%.

When IRGANOR NPA (product name) was added to a palm oil methyl ester blend (POD) (50 ppm anticorrosion/corrosion inhibitor + 15% POD + 85% diesel) a 10–15% increase in brake power was observed.[43]

It was found that the ignition delay of biodiesel could be increased by adding crude glycerine.[44] The potential of diethyl ether, which is a renewable bio-based fuel, as a supplementary oxygenated additive to improve the fuel properties and combustion characteristics of biodiesel KME, such as its high viscosity and cold-start problems, was identified through an experimental investigation.[45]

An experimental investigation was conducted to evaluate the effects of using diethyl ether and ethanol as additives to biodiesel–diesel blends on the combustion characteristics of a DI diesel engine.[46] The test fuels were denoted as B30 (30% biodiesel and 70% diesel by volume), BE-1 (5% diethyl ether, 25% biodiesel and 70% diesel by volume) and BE-2 (5% ethanol, 25% biodiesel and 70% diesel by volume), respectively. The peak pressure rise rate and peak heat release rate of BE-1 were similar to those of B30, and higher than those of BE-2 at lower engine loads. At higher engine loads the peak pressure, peak pressure rise rate and peak heat release rate of BE-1 were the highest and those of B30 were the lowest. BE-1 has better combustion characteristics than BE-2 and B30.

An experimental investigation was conducted to evaluate the effects of using methanol as an additive in biodiesel–diesel blends on the engine performance, emissions and combustion characteristics of a DI diesel engine under variable operating conditions.[47] BD50 (50% biodiesel and 50% diesel by volume) was prepared as the baseline fuel. Methanol was added to BD50 as an additive at 5 and 10% by volume (denoted as BDM5 and BDM10). The results indicate that combustion starts later for BDM5 and BDM10 than for BD50 at low engine loads, but is almost identical at high engine loads. At a low engine load of 1500 rpm, BDM5 and BDM10 show similar peak cylinder pressures and peak pressure rise rates to BD50, and a higher peak heat release rate than that of BD50. At low engine loads of 1800 rpm, the peak cylinder pressure and the peak pressure rise rates of BDM5 and BDM10 are lower than those of BD50, and the peak heat release rate is similar to that of BD50. The crank angles at which the peak values occur are later for BDM5 and BDM10 than for BD50. At high engine loads, the peak cylinder pressures, the peak pressure rise rates and the peak heat release rates of BDM5 and BDM10 were higher than those of BD50, and the crank angle peak values for all tested fuels were almost the same. The power and torque outputs of BDM5 and BDM10 were slightly lower than those of BD50.

7.3   Durability Tests of Diesel Engines Operated using Biodiesel and its Blends

Durability tests are more time-consuming and costly than tests of engine power, economy and emissions. For durability studies, the following aspects were investigated: carbon deposit, engine wear, and problems in the fuel system. Carbon deposits are related to soot formation during combustion of fuel in the engine and fuel oxidation. Researchers investigated the effect of a 20% rice bran oil methyl ester blend with mineral diesel biodiesel on the wear of in-cylinder engine components during 100 h tests.[48] Carbon deposits on the cylinder head, injector tip, and piston crown of the biodiesel engine were significantly lower compared with mineral diesel due to the lower soot formation during combustion of biodiesel. It was also reported that biodiesel improves carbon deposits in the combustion chamber.[49,50]

Fraer *et al.* studied tear-down analysis on a 1996, Mack MR 688 p model vehicle with six cylinders and a compression ratio of 16.5 : 1 which produced 300 horse power at 1950 rpm and was used for postal purposes.[51] The engine and fuel system components were disassembled, inspected and evaluated to compare wear characteristics after four years of operation and more than 6 00 000 miles on B20, no difference in wear or other issues were noted during the engine tear-down. The cylinder heads of the B20 engines contained a large amount of sludge around the rocker assemblies that was not found in the diesel engines. The sludge contained high levels of Na possibly caused by accumulation of soap in the engine oil. The B20 engines required injector nozzle replacement during the evaluation and tear-down period. This was due to the use of out-of-specification fuel. The biological contaminants may have been the cause of filter plugging.

Biodiesel is effective at reducing friction when used as an additive in diesel fuel at low levels.[30,49,50] During tribological investigations of lubricating oils, it was found that the amount of wear debris, soot, resinous compounds, oxidation products, and the moisture content was lower for lubricating oil drawn from the biodiesel-fueled engine compared with the diesel-fueled engine.[49] The improved performance of the biodiesel-fueled system is possibly attributed to the inherent lubricity of biodiesel, resulting in lower wear of vital moving components.

The content of wear metals debris such as Fe, Cu, Al and Pb reduced with increasing addition of palm oil to biodiesel blends, which produced lower wear metal concentrations than ordinary diesel, due to the effect of the corrosion inhibitor in the fuel and lube oil.[50]

7.4 Influence of Biodiesel on Exhaust Emissions from Diesel Engines

Generally, the combustion process in a CI engine occurs only at the interface between the fuel-injection system and air compressed in the cylinder.[52] Thus, oxidation of fuel is not always complete. Incomplete combustion of the fuel produces carbon monoxide (CO), hydrocarbons (HCs), particulate matter (PM) and oxygen-containing compounds such as aldehydes. In addition, the temperature in the cylinder during combustion also promotes the production of NO_x from nitrogen and oxygen in the air.

7.4.1 Particulate Matter, Carbon Monoxide and Hydrocarbon Emissions

Generally there is decrease in PM, CO and HC emissions during use of biodiesel and its blends comparative to neat diesel. In this section we will discuss the various reasons behind this.

Different factors can influence the PM, CO and HC emission characteristics of diesel engines fueled with biodiesel and its blends.

7.4.1.1 Content and Properties of Biodiesel

1. Particulate Matter. Use of pure biodiesel causes a reduction in PM emissions.[3,53–57] Recently, WFOME was tested to check its emissions.[14] The concentration of PM was lower, relative to diesel, by 23–47%. Because of the insignificant sulfur content, sulfur dioxide emissions were lower by 50–100% for different blends.

The emission performance of five pure biodiesels in a Cummins ISBe6 DI engine with a turbocharger and intercooler was studied, and it was found that different biodiesels reduced PM emission by 53–69% on average compared with diesel fuel.[54] A reduction in PM for biodiesels by 75% and 91%, respectively was reported in another study.[105] RME emissions tests showed an increase in PM (10.3%).[58] The results for PM are not significantly different to those of diesel.

Peterson *et al.* compared the effect of blending ratios of 20, 50 and 100% for three kinds of jatropha-, karanja- and polanga-based biodiesels on smoke emissions, and found the use of KB20, KB50 and KB100 caused a reduction in smoke by 28.96, 44.15 and 68.83% with respect to diesel at a rated speed, respectively.[9] Similarly, decreases in smoke for JB20, JB50, JB100, PB20, PB50 and PB100 of 28.57, 40.9, 64.28, 29.22, 44.15 and 69.48% were observed at the rated speed, respectively.

Luj *et al.* demonstrated reductions in PM emissions of 32.3, 42.9 and 53% for B30, B50 and B100 respectively, which were obtained from after-treatment on a HSDI (high-speed direct-injection) four-cylinder, 1.6 L, turbo diesel engine.[59]

However, several authors have reported a reverse change with increasing proportion of pure biodiesel. In one such study, the effect of blends of 10, 20 and 50% biodiesel from sunflower and olive oil on a single cylinder, indirect-injection (IDI), stationary diesel engine were investigated and maximum PM emissions were observed for the 10% blends and minimum PM emissions for the 50% blends at the different loads.[60]

Many authors have attributed the reduce in PM emissions to the higher oxygen content in biodiesel, which causes more complete combustion, and further promotes the oxidation of soot.[61] Scientists have investigated the effect of fuel cetane number (CN) and aromatics on the combustion process and emissions of a DI diesel engine, and reported that PM emissions increased at high loads when the aromatic content was increased with constant CN.[62] As for the effect of the CN of biodiesel, it was further shown that reducing the CN resulted in a decrease in PM at high loads.

The higher density and viscosity of biodiesels could affect the volatilization and atomization processes, and further deteriorate combustion in the chamber.[16]

Biodiesels which had the effect of reducing PM, in descending order, were WCOME, palm oil methyl ester (PME), CSOME, RME and soybean methyl ester (SME). This is due to the combination of different oxygen contents, viscosities and CNs.[54]

2. Carbon Monoxide. CO emissions reduce when pure biodiesel is used as a diesel engine fuel.[3,16,17,21,26,34,53–57] Experiments have been carried out to estimate the emission characteristics of a single cylinder; four-stroke variable compression ratio multi-fuel engine fuelled with WCOME and its blends with standard diesel.[12] Tests were conducted using fuel blends of 20, 40, 60 and 80% biodiesel with standard diesel, with an engine speed of 1500 rpm, a fixed compression ratio of 21 and at different loading conditions. The use of the blends resulted in a reduction in CO emission.

With increasing biodiesel content in the blends, CO emissions reduce due to an increase in oxygen content.[16,21,36,63] It was found that at full load, the CO emissions of diesel were the highest (15.2 g kWh^{-1}), with the other fuels recording lower emissions: BD10 = 12.8 g kWh^{-1}, BD30 = 11.7 g kWh^{-1}, BD50 = 10.7 g kWh^{-1} and BD100 = 11.4 g kWh^{-1}.[63] It has also been reported that there is a variation in CO emissions of blends with increasing biodiesel content.[16] Furthermore, it was also found that, with the increase in biodiesel percentage in the blends, there was no obvious difference in the CO emissions at partial loads, but they fluctuated at full loads, due to the interaction of low-volatility polymers and the higher oxygen content.[64]

Recently, WFOME was tested as a diesel fuel substitute.[14] CO emissions were reduced by 21–45% for different blends.

The feedstock of biodiesel may affect CO emissions. Wu *et al.* studied the difference in CO emissions for five biodiesels (CSOME, SME, RME, PME and WCOME).[54] They attributed the difference in CO emissions for the five types of biodiesel and diesel fuels at high loads to the oxygen content, but at low loads to the CN, and concluded that CO emissions decreased consistently for both biodiesels and diesel fuel as CN increased. Biodiesel has a higher CN, which results in a lower possibility of the formation of a fuel-rich zone and thus reduces CO emissions.

Another experiment was carried out for esterified karanja oil (B100) and its blends.[65] The minimum and maximum CO produced was 0.004 and 0.016%, respectively resulting in a reduction of 94 and 73% for B20 and B100 as compared to diesel.

Methyl esters emitted less CO compared to ethyl esters.[66] Knothe *et al.* tested an engine with lauric (C12 : 0), palmitic (C16 : 0) and oleic (C18 : 1) methyl esters, and reported that CO emissions reduced with increasing chain length.[67]

3. Hydrocarbons. HC emissions reduce when pure biodiesel is used instead of diesel in diesel engines.[17,53,54,60] Tests have been conducted using fuel blends of 20, 40, 60 and 80% biodiesel with standard diesel, with an engine speed of 1500 rpm, a fixed compression ratio of 21 and at different loading conditions. Use of the blends resulted in a reduction of HC emissions.[12]

Many researchers have agreed that HC emissions decrease with increasing biodiesel percentage in the blend.[23,36,64,60] It was found that the reduction in HC emissions was linear with the addition of biodiesel to the blends.[23] In contrast, Song *et al.* pointed out that a greater reduction in CO emissions appears with low biodiesel content, that is, lower biodiesel concentrations are more effective than higher ones.[64]

It was found that HC emissions could be reduced by 22.9, 17.7 and 16.4% for B30, B50 and B100 respectively, compared with diesel. Luj *et al.* explained that the lower heating values of pure biodiesel imply higher fuel consumption and therefore high local fuel-to-air ratios which cause an increase in HC emissions. A high catalyst efficiency reduces the advantages of biodiesel in terms of HC emissions.[59]

Some studies have shown that the source of biodiesel has an effect on HC emissions, for example it has been reported that the five typical methyl ester biodiesels reduced HC emissions by 45–67% on average, as mentioned above, and attributed the difference in HC emissions with different biodiesels to a combined effect of oxygen content and CN.[54] The properties of biodiesel are related to HC emissions for instance, it has been proved that the increase in chain length or saturation level of several biodiesels leads to a higher reduction in HC emissions on an 11.1 L engine.[68] Similarly, it has been reported that, HC emissions reduced with increasing in chain length when researchers tested lauric (C12 : 0), palmitic (C16 : 0) and oleic (C18 : 1) methyl esters on a six-cylinder engine, and the HC emissions reduced by 50% for pure biodiesel relative to diesel.[67] It was explained that the decrease in HC emissions was caused not only by the oxygen content but also by the CN.[54] The higher CN of biodiesel could reduce the burning delay, which would result in a reduction in HC emissions.[69,70]

7.4.1.2 Engine Load

1. Particulate Matter. Engine load plays a significant role in the PM emissions of biodiesel. Researchers tested mahua biodiesel and its blends at different loads on a single-cylinder, four-stroke, WC Ricardo E6 engine, and found that the smoke level increased sharply with the increase in load for all fuels tested.[25] They explained that this was mainly due to the decreased air-to-fuel ratio at higher loads when larger quantities of fuel are injected in to the combustion chamber, much of which goes unburnt into the exhaust.

It was also reported that reductions in PM emissions themselves reduced at low and middle loads. Many authors in this area have agreed on this trend of PM emissions with load.[71] However, some researchers have observed the reverse trend.[72–74] These authors have explained that, this trend is because particles are mainly formed during diffusion combustion, and most of the combustion is diffusive at high loads, which means that the oxygen content of biodiesel is more effective in reducing PM emissions.

2. Carbon Monoxide. The literature reports that CO emissions increase with increasing engine load.[22,26,55] The main reason for this increase is because the air-to-fuel ratio decreases with increasing load, which is typical of all internal combustion engines. In contrast, it was reported that CO emissions reduced with increasing load, because the increase in combustion temperature lead to more complete combustion during higher loads.[57,75,76] It was also concluded that CO emissions decrease as the load increases, but that they increase slightly at heavy or full loads. Other researchers have found that CO emissions were lower at intermediate loads, but higher at low or no loads, heavy loads and full loads.[63]

3. Hydrocarbons. The effect of engine load on HC emissions for biodiesel has been initially studied, but the conclusions reached were inconsistent. Researchers showed experimentally an increase in HC emissions with load increase due to high fuel consumption at high loads.[26] However, others found that the HC emissions for biodiesel reduce as the load increases.[57,64,71,75]

7.4.1.3 Engine Speed

1. Particulate Matter. The higher the engine speed, the lower PM emissions are.[55,56,64] This is because of improved combustion efficiency which can be attributed to an increase in turbulence effects with an increase in engine speed, enhancing the extent of complete combustion. However, Utlu and Kocak reported that the impact of engine speed appeared to fluctuate as PM emissions reduced at low speed, and increased in the range 2000–4000 rpm, then decreased again above 4000 rpm.[3]

2. Carbon Monoxide and Hydrocarbons. These emissions decrease for biodiesel with increasing engine speed as result of better air–fuel mixing processes and an increase in the fuel-to-air equivalence ratio.[77–79]

7.4.1.4 Injection Timing

1. Particulate Matter. Usually, the start of injection of biodiesel occurs earlier than for diesel due to its higher density and viscosity and lower compressibility.[53,57,71] Therefore, researchers have studied the effect of injection timing and showed that the smoke emission with biodiesel generally increased when the injection timing was retarded.[34,36] Initially the smoke level of reference diesel falls when the injection timing is advanced to 23° from 19° before top dead center (BTDC) and then increases when the injection timing is advanced further.

Recently, an investigation of the impact of injection strategy and biodiesel on engine PM emissions with a common-rail turbocharged DI diesel engine was conducted at moderate speed and different load/torque conditions.[80] For PM emissions, it was found that an increase of fuel injection pressure decreases PM emissions at all load conditions. Meanwhile, biodiesel fuel has a more significant effect on PM emissions at low loads and a less significant effect at moderate-to-high loads.

2. Carbon Monoxide and Hydrocarbons. It has been reported that CO and HC emissions reduce when the injection timing is advanced for biodiesel fuels.[34,81] Researchers tested the effect of the use of retarded ignition timing (by 3 °CA) on the emissions of biodiesel from rapeseed oil, and found that the retardation resulted in increased CO and HC emissions.[36]

7.4.1.5 Exhaust Gas Recirculation

It was found that smoke emission was increased for biodiesel blends with exhaust gas recirculation (EGR) from 0 to 20%, although the smoke levels were generally lower and remained at considerably lower values.[36] Researchers attributed this to the decreased availability of oxygen for the combustion of fuel, which results in relatively incomplete combustion and increased formation of PM. Additionally, researchers investigated the effect of EGR in the range 0 to 100%.[82] They found that there are two distinct relationships which describe the effect of EGR on soot. In the first relationship, the soot increases with increasing EGR, up to 50–70%. After this point, the soot decreases with increasing EGR.

7.4.1.6 *Additives*

1. Particulate Matter. Oxygenates such as ethanol, methanol and other alcohols were added to biodiesel and they all caused a further decrease in PM emissions due to the enrichment of the oxygen content of the fuel.[71,75] It was found that biodiesel blends with Mg and Mo additives demonstrated a greater reduction in PM emissions due to the catalyst effect of the metals.[37,38] But other researchers found no difference in exhaust emissions between biodiesel fuels with or without antioxidants. Antioxidants also have a significant influence on fuel consumption.[83]

Recently, another study focused on the effect of blending an oxygenated component (diglyme) and biodiesel fuel with conventional diesel.[84] The basis of comparison was around the oxygen content of the blends paying significant attention to PM. In assessing the potential of the two blends, the emission performances were analyzed and directly compared. The addition of diglyme resulted in a reduction in PM emissions.

Recently, experiments were conducted on a four-cylinder DI diesel engine using an ultra-low-sulfur diesel (ULSD) blended with ethanol, biodiesel and diglyme to investigate the PM emissions of the engine under five engine loads at an engine speed of 1800 rpm.[85] Four diesel–ethanol blends with oxygen concentrations of 2, 4, 6 and 8%, five diesel–biodiesel blends and five diesel–diglyme blends with oxygen concentrations of 2, 4, 6, 8 and 10% were studied. The results showed that PM emissions decrease with increasing oxygenate content in the blends. With the same oxygen content, different oxygenates had different influences on emissions, particularly on PM reduction efficiency, indicating that physical properties and chemical structure also influence PM emissions.

2. Carbon Monoxide and Hydrocarbons. A decrease in CO and HC emissions of biodiesel containing metal-based additives has been reported. Researchers compared the CO emissions of B20, B20X (1% 4-NPAA added to B20) and B0, and found that the B20X fuel produced the lowest level of CO emissions, which was 0.1%, followed by B20 (0.2%) and B0 (0.35%).[30] Others observed that the CO emissions of biodiesel fuel decreased with Mg- and Mo-based additives.[37] Cheung *et al.* tested the exhaust emissions of pure biodiesel and its blends with 5% (BM5), 10% (BM10) and 15% (BM15) methanol.[75] It was found that the HC emissions of BM5 and BM10 were lower than those of biodiesel, except at the lowest engine load of 0.08 MPa while the HC emissions of BM15 were higher than those of biodiesel except at the highest engine load of 0.70 MPa. It was explained that the small amount of methanol in BM5 could increase the oxygen content of the blended fuel and reduce the viscosity and density of the blend so that HC emissions were reduced. But the cooling effect of methanol for BM15 dominates the increase of HC emissions. For BM5, the CO emissions were even lower than those of biodiesel, with a reduction of 6% on average, based on different engine loads. However, the CO emissions of BM10 and BM15 were higher than those of biodiesel at light and medium engine loads, while lower than those of biodiesel at high engine loads.

Decreased CO and HC emissions were reported when an engine was operated with mahua biodiesel blended with ethanol.[86] Furthermore, it was reported that the reduction in CO emission levels with the addition of oxygenates (ethanol) was obvious when performance and emissions were compared with mahua methyl ester (MME), MME E20, MME E10 and MME E10 D10 (MME with 10% diesel and 10% ethanol). BD15E5 mixed fuel (15% biodiesel, 5% bioethanol and 80% diesel) yielded lower HC emissions than that of B20 (20% biodiesel and 80% diesel) in a common-rail direct-injection diesel engine.[87]

An experimental investigation was conducted to evaluate the effects of using methanol as an additive to biodiesel–diesel blends on the engine emissions characteristics of a DI diesel engine under variable operating conditions.[47] BD50 (50% biodiesel and 50% diesel by volume) was prepared as the baseline fuel. Methanol was added to BD50 as an additive by 5 and 10% by volume (denoted as BDM5 and BDM10). CO emissions were slightly lower, and HC emissions were almost similar to those of BD50 at full engine load.

7.4.1.7 *Analysis of Aforementioned Discussion*

Based on analysis above, the following conclusions can be drawn:

(a) PM, CO and HC emissions reduce when pure biodiesel is used instead of diesel.
(b) PM, CO and HC emissions for biodiesel reduce with increasing biodiesel content. The feedstock of biodiesel and its properties have an effect on PM, CO and HC emissions, especially the chain length and saturation level.
(c) Most researchers have found that the there is an increase in PM, CO and HC emissions with increasing engine load.
(d) The higher the engine speed, the lower the PM, CO and HC emissions.
(e) The advance in injection timing of biodiesel favors lower PM, CO and HC emissions.

7.4.2 NO_x Emissions

NO_x emission is discussed separately because most researchers believe that these emissions increase with the use of pure biodiesel, therefore in this section we will discuss the various reasons for this.

Different factors can influence the NO_x emission characteristics of a diesel engine fueled with biodiesel and its blends.

7.4.2.1   Content and Properties of Biodiesel

Most researchers believe that NO_x emissions increase with the use of pure biodiesel.[15,23,31,34,53–57,63,64,75,86,87] However, there are a few reports which show that NO_x emissions decrease with use of pure biodiesel.[3,16,35,60,66,77]

The literature also shows that NO_x emissions increase with the increase in content of biodiesel in blends.[4,23,24,28,59,64,87] For example, researchers tested a HSDI, four-cylinder, 1.6 L, turbocharged (TU) diesel engine fueled by biodiesel and its blends B30, B50 and B100 and observed that the increase in NO_x emissions for B30, B50 and B100 was 20.6, 25.9 and 44.8%, respectively.[59] Experiments with WCOME blends of 20, 40, 60 and 80% resulted in an increase in NO_x emissions.[12]

Other researchers investigated the engine performance and emissions of CSOME and its blends (B5, B20, B50 and B75) on a single-cylinder DI, AC diesel engine.[16] It was observed that, in general, increasing the content of biodiesel in the blends resulted in a reduction of NO_x emissions, but use of B5 resulted in a decrease in the NO_x emissions.

Ulusoy *et al.* showed that there is no regular change in NO_x emissions with increasing content of pure biodiesel.[55] It was found that the B35 blend (4.075% oxygen) produced maximum NO_x emissions compared to all other blends, including pure biodiesel RME (10.9% oxygen). They suggested that this resulted from some indigenous features of the fuel, *i.e.*, not containing any aromatic compounds, its slower evaporation and its lower heating value.

It was found that the NO_x concentration in exhaust emissions increases with increasing biodiesel density, iodine value and percentage of unsaturation.[88] Increasing the density may increase NO_x emissions because the fuel injector injects a constant volume, but a larger mass, of the more dense fuels. When a larger mass of fuel is burned, more NO_x is produced. Secondly, a higher density will result in a higher bulk modulus, advancing the effective injection timing and thereby causing NO_x emissions to increase. The injection advance results in a longer ignition delay since the fuel is injected in air at lower temperatures and pressures. Thirdly, the iodine value, which is a direct measure of unsaturation is highly inversely correlated with CN. Thus, excessive ignition delay and poor combustion performance may also cause higher NO_x emissions.

Other researchers have observed that the higher CN of biodiesel shortens the ignition delay, and thus combustion advances, increasing NO_x emissions.[21,54,71] However, there may be objections to this reasoning. A higher CN will not only lead to earlier burning but also to lower pre-mixed combustion, which will lead to softer changes in pressure and temperature, thus causing less NO_x formation, not more.

The higher oxygen content in biodiesel enhances the formation of NO_x.[55] Results have shown that maximum NO_x emissions increase proportionally with the mass percent of oxygen in RME–diesel blends. However, Lapuerta *et al.* have suggested that the oxygen content in biodiesel has no influence on NO_x emissions.[89] Alam *et al.* found that there is no significant difference in the amount of oxygen in the exhaust between the fuels, No. 2 diesel fuel (no oxygen), No. 1 diesel fuel (no oxygen), SME (10.97% oxygen by mass) and its 20% blend. The NO_x emissions of the SME and 20% blend were increased by 11.2 and 0.6%, respectively, compared to the No. 2 diesel, but the NO_x emissions were reduced by 6% for No. 1 diesel fuel compared to No. 2 diesel fuel.[90] Therefore, they suggested that more research is required regarding the other properties of biodiesel and their effects on combustion and the fuel system to better explain NO_x emission increases.

Results have also shown that NO_x emissions increase because the average carbon chain length lowers and the number of unsaturated compounds increases.[17,68] The different air-to-fuel equivalence ratios, residual amounts of methanol, and the injection and ignition delays may also contribute to different NO_x emissions.[53]

7.4.2.2   Engine Load

NO_x emissions increase as the engine load increases as a result of the higher combustion temperatures present at higher engine loads.[23–26,55,64,66,71,75] As the load increases, the overall fuel-to-air ratio increases, resulting in an increase in the average gas temperature in the combustion chamber and therefore NO_x formation which is sensitive to temperature increases.[23,24,35] However, it was also found that NO_x emissions increase at light loads, although NO_x emission increases at middle and high loads may also be due to timing changes made by the light-load advance mechanism on the fuel-injection pump.[57]

Murillo *et al.* found that NO_x emissions decreased as the load increased on a single-cylinder, four-stroke, NA, DI diesel outboard engine during an ISO C-3 test cycle.[63] They explained that this is probably due to the increase in turbulence inside the cylinder, which may contribute to faster combustion and a lower residence time of the species in the high-temperature zones.

7.4.2.3   Engine Speed

Some researchers found that NO_x emissions reduced with increasing engine speed due to the shorter residence time available for NO_x formation, which may be the result of an increases both in the volumetric efficiency and flow velocity of the reactant mixture at higher engine speeds.[31,56]

It was also reported that NO_x emissions increased at light loads, and reached a maximum value at medium loads, then reduced with increasing engine speed when a B60 biodiesel blend was used at full load on a single-cylinder, four-stroke, AC, DI diesel engine.[37] Usta observed different effects of engine speed on NO_x emissions under different load conditions without explanation, that is, as the engine speed is increased, the NO_x emissions increased at full load, and slightly increased at 75% load, but gradually reduced at 50% load.[78] Others found that an increase in NO_x was observed between maximum torque and maximum power speeds for WFOME and for a reference diesel fuel, depending on the exhaust temperatures and increasing volumetric efficiencies.[3]

7.4.2.4   Injection Timing

Hirkude and Padalkar found that NO_x emissions increased as the injection timing was reduced.[14] But others found that a decrease in injection timing resulted in reduced NO_x emissions.[36]

Recently, an investigation of the impact of injection strategy and biodiesel on engine NO_x emissions with a common-rail turbocharged DI diesel engine was conducted at moderate speed and different load/torque conditions.[80] The fuels included a baseline ULSD and a B40 (v/v) blend of an SME-based biodiesel and ULSD. Single fuel injections at the start of injection timings over a range of 9° before to 3° after top dead center with different fuel injection pressures were investigated. It is found that at all load conditions, an increase of fuel injection pressure significantly increases NO_x emissions, and that with the same injection strategy as the baseline diesel, biodiesel fueling increases NO_x emissions. Linear correlations between the average oxygen equivalence ratio of the fuel–air mixture at the auto-ignition zone near the lift-off length and brake-specific NO_x emissions were observed for all load conditions, regardless of fuel type. This confirms that the dominant factor that determines NO_x emissions is the ignition event controlled by the oxygen equivalence ratio at the auto-ignition zone.

7.4.2.5   Exhaust Gas Recirculation

Zheng *et al.* investigated the effect of EGR on a single-cylinder, four-stroke, NA, DI diesel engine, and found that there were slight differences in NO_x emissions between biodiesels and diesel, but NO_x emissions for all fuels decreased with increasing EGR.[82] Researchers operated a single-cylinder engine with different EGR ratios and found that, there was no significant difference between diesel and neem biodiesel with a 5–30% EGR ratio, although they observed an increase in NO_x emissions without EGR.[91]

7.4.2.6   Additives

Methanol and ethanol have been added in small amounts to improve NO_x emissions for biodiesel.[4,75,86] For example, it was found that ethanol could act as an effective NO_x emissions reducing additive, because the addition of only 5% ethanol to biodiesel in a BE5 experimental fuel drastically suppressed the increase (2.6%) in NO_x compared to B100 (12% increase).[4]

Recently, another study focused on the effect of blending an oxygenated component (diglyme) and biodiesel fuel with conventional diesel.[84] In assessing the potential of the two blends, the combustion and emission performances were analyzed and directly compared. The addition of diglyme resulted in a reduction in NO_x emissions, which resulted in control of the overall NO_x trade-off.

Experiments by Wang *et al.* were conducted on a four-cylinder DI diesel engine using ULSD blended with ethanol, biodiesel and diglyme to investigate NO_x emissions.[85] The results showed that the addition of ethanol to diesel fuel increases NO_x (NO_2) emissions. Addition of diglyme to diesel fuel can simultaneously decrease both PM and NO_x emissions.

It was found that NO_x emissions were most effectively reduced by burning an O–W–O three-phase biodiesel emulsion that contained aqueous ammonia, particularly at lower engine speeds on a four-stroke, four-cylinder, NA, DI diesel engine.[31] Engine emissions were also studied with a synthetic Mg additive on a single-cylinder, DI diesel engine.[38] It was reported that NO_x emissions increased by 5% for a blend of 10% biodiesel–diesel at full load and different engine speeds from 1800 to 3000 rpm. Keskin *et al.* studied the effect of Mg and Mo as combustion catalysts on engine performance and emissions for B60 biodiesel blends on a single-cylinder, four-stroke, AC, DI diesel engine, and found that lower NO_x emissions were measured with Mg at low engine speeds and with Mo at higher speeds.[37] It was reported that NO_x emissions can be reduced by operating an engine in dual-fuel mode *i.e.*, by using a fuel with a high octane number as the primary fuel and one with a high CN as the pilot fuel.[77]

7.5 CO$_2$ Emissions

It has been reported that the use of biodiesel results in fewer carbon dioxide (CO_2) emissions than diesel during complete combustion due to the lower carbon-to-hydrogen ratio.[3,31,53] This is attributed to the fact that biodiesel is a low-carbon fuel, so has a lower elemental carbon-to-hydrogen ratio than diesel. However, it has also been reported that CO_2 emissions rise or are similar, due to more efficient combustion.[22,55] Biodiesel can cause a 50–80% reduction in CO_2 emissions compared to petroleum diesel.[15,71] Experiments with WCOME blends of 20, 40, 60 and 80% resulted in a decrease of CO_2 emissions.[12]

7.6 Non-Regulated Emissions

There has been increasing interest recently in the non-regulated emissions such as carbonyl, aromatic and polyaromatic compounds from biodiesel.

7.6.1 Carbonyl Compounds

Different factors can influence carbonyl compounds emission characteristics of diesel engine fueled with biodiesel and its blends.

7.6.1.1 Content and Properties of Biodiesel

It has been found that carbonyl compound emission increases when using pure biodiesel or its blends because of their higher oxygen content.[75,92–94] Fontaras *et al.* found 13 carbonyl compounds in exhaust gases and measured their concentrations over various driving cycles with a B100 biodiesel and petroleum diesel and the results demonstrated a significant increase of carbonyl emissions with the use of pure biodiesel, probably due to the oxygen atoms in the ester molecules.[95]

He *et al.* analyzed 14 carbonyl compound emissions, mainly including formaldehyde, acetaldehyde, acrolein, acetone *etc.*, and found that biodiesel-fueled engines had almost triple the carbonyl emissions of diesel-fueled engines and emitted a comparatively high content of propionaldehyde and methacrolein.[94] It was found that, when a 20% blend fuel from rapeseed oil biodiesel was tested in a six-cylinder engine, there was a marked increase in formaldehyde emission, but no significant differences in acrolein, acetaldehyde and propionaldehyde emissions.[96]

However, a reduction in carbonyl emissions with pure biodiesel and a smaller reduction with a B20 blend has also been reported.[97] Peng *et al.* found that B20 (20% waste cooking oil biodiesel and 80% diesel) generated slightly fewer emissions in total aldehyde compounds than diesel.[98]

It was found that the quality of biodiesel w.r.t. the fatty acid profile, iodine number, and purity level plays a role in the formation of certain carbonyl emissions.[95] Arapaki *et al.* reported that the acrolein concentration in the emissions was strongly related to the higher glycerine content of biodiesel used.[99] The presence of short-chain chemicals in biodiesel favors the formation of formaldehyde during combustion.[100] All carbonyl emissions exhibited a strong correlation (correlation coefficients better than 0.96) with biodiesel content, which indicates that carbonyl emissions are strongly influenced by the biodiesel content and that the biodiesel ester molecules are probably the source of these carbonyls. However, Liu *et al.* reported that the total concentration of emitted carbonyls did not increase with biodiesel content.[93]

Costa Neto *et al.* presented the utilization of used frying oil for the production of biodiesel, and discussed the performance of biodiesel in diesel engines.[101] The emissions derived from engines running on unused vegetable oils were compared to those derived from engines running on conventional diesel. Two nitro-group-containing polyaromatic hydrocarbons (PAHs) were identified, namely 2-nitrofluorene (2-NFLU) and 1-nitropyrene (1-NP). The mean values for their sum, in emissions from engines using diesel and B20 were, respectively, 156 ± 22 and 248 ± 169 ng kWh^{-1}, with NFLU predominating. Regarding the carbonyl compounds, the sums of the 12 compounds found in this work, were 79.3 and 94.5 mg kWh^{-1} for diesel and B20 respectively. Formaldehyde, followed by acetaldehyde, were the two most abundant in both fuels, with average values of 42.7 and 15.8 mg kWh^{-1} in diesel and 50.6 and 16.8 mg kWh^{-1} in B20. The results of this study point to an increase in carbonyl compound emissions.

7.6.1.2   Engine Load and Speed

Carbonyl compound emissions increased when the engine was run on biodiesel and its blends at 10, 33, and 55%.[93] Formaldehyde emissions increased when the engine load was increased from 0.08 to 0.38 MPa, but decreased when the engine load was increased from 0.38 to 0.70 MPa.[75] Formaldehyde emissions increased with engine load under medium and high engine loads at an engine speed of 1200 rpm, and decreased with engine load under medium and high engine loads at an engine speed of 1400 rpm.[102]

7.6.1.3   Additives

Formaldehyde and acetaldehyde emissions increased with increasing methanol fraction in a biodiesel blend fuel compared with diesel fuel.[75] Formaldehyde emissions increased with the methanol fraction, and the authors concluded that exhaust formaldehyde was mainly produced from the methanol.[103] However, it has also been reported that acetaldehyde emission was a result of ethanol or straight-chain hydrocarbons, and that methanol had no significantly effect on its emission. Arapaki *et al.* reported that methanol itself has no effect on acetaldehyde emissions, however, the addition of methanol to biodiesel could increase the oxygen content leading to an increase of acetaldehyde emissions, and simultaneously the contents of straight chain-hydrocarbons will change after the addition of methanol.[99] Therefore, methanol tends to increase acetaldehyde emissions.

7.6.2   Aromatic and Polyaromatic Compounds

Different factors can influence the aromatic and polyaromatic compound emission characteristics of a diesel engine fueled with biodiesel and its blends.

7.6.2.1   Content and Properties of Biodiesel

Aromatic compounds and their derivatives are toxic, mutagenic and carcinogenic. A decrease in the aromatic and polyaromatic emissions of biodiesel compared to diesel has been reported.[75,92,94,96,97] For example, researchers found that PAH and nitro-PAH emission reduced by 50–75% on three different engines fueled by biodiesel.[97] Turrio-Baldassarri *et al.* explained that the reduction in PAH emission was usually due to the enhanced absorption of PM to these components.[96]

However, Munack *et al.* demonstrated that the increase of rapeseed biodiesel content in a blend fuel caused an increase in benzene emissions on a single-cylinder 4.2 kW engine.[104] They also tested the blend on a 52 kW engine and observed that aromatic emissions increased slightly. Biodiesel blends with diesel are expected to emit fewer light aromatic compounds than pure diesel, since in vegetable oils there is virtually no aromatic content.[99]

Researchers have observed this in their work, with mean values of 29.9 and 13.9 mg kWh^{-1} for diesel and B20, respectively, benzene and toluene comprising more than 60% of the total.

7.6.2.2 Engine Load and Speed

Researchers measured the benzene, toluene and xylene (BTX) emissions on a diesel engine under five engine loads at a steady speed of 1800rpm with Euro V diesel fuel, pure biodiesel and biodiesel blends with 5, 10 and 15% methanol.[75] They observed that the BTX emissions decreased with engine load.

7.6.2.3 Additives

Compared with diesel, the BTX emissions of biodiesel were lower, and with increasing percentage of methanol in the blends, benzene emissions decreased due to an increase of oxygen in the biodiesel which improves combustion and promotes the degradation of benzene.[75]

7.7 Statistical Relationship between Biodiesel Performance and Emission Characteristics with Fatty Acid Methyl Ester Composition

The effect of the unsaturated fatty acid methyl ester (FAME) composition of biodiesel on the combustion, performance and emissions characteristics of a diesel engine was investigated and correlated.[88] For this experiment, 13 different biodiesel fuels with different fatty acid compositions were selected. Performance and emissions tests on a single-cylinder DI diesel engine were conducted using these biodiesel fuels. The researchers also established the statistical relationship between performance, combustion and emission characteristics and FAME content.

7.7.1 Correlation of Combustion Parameters

It has been proved that the fuel dynamic injection timing is positively correlated with the percentage of unsaturation (X) and the density.[88] That is, the fuel-injection timing is faster for higher density fuels. The fuel-injection timing is mainly influenced by the fuel properties, such as its bulk modulus and viscosity. The higher the bulk modulus and the viscosity, the faster the injection timing is. The bulk modulus of unsaturated FAMEs is higher than that of saturated FAMEs and increases with increasing density. Eqn (1) shows the variation of ignition delay (ID) with X:

$$ID = 0.0962X + 4.6837 \quad \left(R^2 = 0.898 \right) \tag{1}$$

where for every 1% increase in X, an increase of 0.0962 units (in terms of degree crank angle) in ID is observed.

There is only poor correlation between dynamic injection (DI) and X [eqn (2)]:

$$DI = 0.0438X + 12.655 \quad (R^2 = 0.129) \tag{2}$$

It was found to be highly complex to relate the FAME content and properties of biodiesel with the heat release rate with any precision. It was found that the maximum heat release rate (MHR) is negatively correlated with X, however the correlation coefficients are not very significant [eqn (3)]:

$$MHR = -0.062X + 76.65 \quad (R^2 = 0.032) \tag{3}$$

It was also found that the peak cylinder pressure (PCP) is negatively correlated with X, however, the correlation coefficients are also poor [eqn (4)]:

$$PCP = -0.0779X + 71.898 \quad (R^2 = 0.427) \tag{4}$$

The variation of total combustion duration (TCD) with X is shown in eqn (5):

$$TCD = -0.2806X + 72.202 \quad (R^2 = 0.939) \tag{5}$$

From eqn (5), it can be proposed that every 1% increase in X may result in a decrease of 0.2806 units (°CA) in TCD.

Eqn (6) shows the variation of the cumulative heat release (CHR) with X:

$$CHR = -2.7572I + 1269.4 \quad (R^2 = 0.0801) \tag{6}$$

The correlation coefficient is very poor.

7.7.2   Correlation of Performance Parameters

Eqn (7) shows the variation of the BSFC with X with good R^2:

$$BSFC = 0.0005X + 0.2921 \quad (R^2 = 0.880) \tag{7}$$

From eqn (7), it can be concluded that every 1% increase in unsaturation may result in an increase of 0.0005 units (g kWh^{-1}) in $BSFC$. It can be stated that the BSFC increases with increasing percentage of unsaturation, since an increase in X will result in an increase in density and in a decrease in heating value.

Eqn (8) shows the variation of the BSEC with X with good R^2:

$$BSEC = 0.011X + 12.224 \quad (R^2 = 0.884) \tag{8}$$

From the slope of the equation $y = 0.011x + 12.224$, the gradient between the BSEC and the percentage of unsaturation can be proposed as 0.01. Every 1% increase in unsaturation may therefore result in an increase of 0.011 units (MJ kWh^{-1}) in the BSEC.

The BTE is highly negatively correlated with X [eqn (9)]:

$$BTE = -0.0236X + 29.363 \quad (R^2 = 0.886) \tag{9}$$

The exhaust gas temperature (EGT) is also highly correlated with X [eqn (10)]:

$$EGT = 0.823X + 293.21 \quad (R^2 = 0.976) \tag{10}$$

7.7.3 Correlation of Emission Parameters

Eqn (11) shows the highly correlated variation of NO_x emissions with X:

$$NO_x = 0.0666X + 8.5833 \quad (R^2 = 0.972) \tag{11}$$

An increase of 0.0666 units (g kWh^{-1}) in NO_x might therefore be expected for every 1% increase in unsaturation.

There is only a poor correlation between CO emissions and X [eqn(12)]:

$$CO = 0.0401X - 0.2245 \quad (R^2 = 0.667) \tag{12}$$

From eqn (12), the probable increase in CO emissions may be proposed as 0.0401 units (g kWh^{-1}) for every 1% increase in unsaturation.

There is also only a poor between HC emissions and X [eqn(13)]:

$$HC = 0.002X + 0.2001 \quad (R^2 = 0.597) \tag{13}$$

This correlation analysis proves that CO and HC emissions are positively correlated with the percentage of unsaturation. That is CO and HC emissions increase with increasing unsaturation. This can be explained due to the lower oxygen concentration in more highly unsaturated biodiesel fuels. The oxygen content decreases with increasing unsaturation. CO is believed to be formed at the borders between the lean flame out region (LFOR) and the lean flame region (LFR) during the early stages of spray combustion.[88] The LFOR is the region that is nearer to the leading edge of the spray (downwind). In this region the mixture is too lean to ignite or support combustion. At this stage, the primary reaction can take place and the initial hydrocarbons may reduce to CO, H_2, and H_2O. As the local temperature is not high enough at this stage, very few oxidation reactions take place. An increase in unsaturation may tend to decrease the gas temperature, and a reduced gas temperature may not provide a positive situation for complete oxidation, thereby increase the CO concentration.

On the other hand, the LFOR is one of the main contributors to the HC concentration in the exhaust. If the ignition delay is larger, the droplets and vapor will be carried farther away from the center line of the spray in the downward direction, resulting in a wider LFOR. If unsaturation increases, the ignition delay also increases which results in an increased width of the LFOR. An increase in the width of the LFOR eventually increases the HC concentration in the exhaust.

Eqn (14) shows the variation of smoke with X:

$$S = -0.0163X + 2.2909 \quad \left(R^2 = 0.786\right) \tag{14}$$

The gradient between the amount of smoke and the percentage of unsaturation is -0.0163 It is therefore proposed that every 1% increase in unsaturation may result in a decrease of 0.0163 units (Bosch Smoke Unit (BSU)) in smoke.

References

1. P. K. Sahoo, L. M. Das, M. K. G. Babu and S. N. Naik, *Fuel*, 2007, **86**, 448.
2. S. Puhan, N. Vedaraman, B. V. B. Ram, G. Sankarnaryanan and K. Jeychandran, *Biomass Bioenergy*, 2005, **28**, 87.
3. Z. Utlu and M. S. Kocak, *Renewable Energy*, 2008, **33**, 1936.
4. A. C. Hansen, M. R. Gratton and W. Yuan, *Trans. ASABE*, 2006, **49**, 589.
5. H. S. Yesu and I. Cumali, *Energy Sources, Part A*, 2006, **28**, 389.
6. N. Usta, *Biomass Bioenergy*, 2005, **28**, 77.
7. J. G. Suryawanshi and N. V. Deshpande, ASME [ICEF] 2004, 866. Conference paper, long beach, CA, USA, ASME, New York USA.
8. K. M. Senthil, A. Ramesh, B. Nagalingam and K. V. Gopalakrishnan, *Proceedings of the 16th National Conference on IC Engines and Combustion*, ed. P. K. Bose, Narosa Publishing House, New Delhi, 2000, p. 89.
9. C. L. Peterson, D. L. Reece, J. C. Thompson, S. M. Beck, C. Chase, *Biomass Bioenergy*, 1996, **10**, 331.
10. H. Raheman and A. G. Phadatare, *Biomass Bioenergy*, 2004, **27**, 393.
11. R. Kadiyala, B. V. Apparao, S. Chandra Prasad and S. Niranjan Kumar, *Proceedings of the 19th National Conference on IC Engines and Combustion*, eds. R. Kadiyala, B. V. Apparao, S. Chandra Prasad and

S. Niranjan Kumarm, Annamalai University, Chidambaram, India, 2005, p. 47.

12. K. Muralidharan, D. Vasudevan and K. N. Sheeba, *Energy*, 2011, **36**, 5385.

13. Y. Haik, Y. E. Selim Mohamed and T. Abdulrehman, *Energy*, 2011, **36**, 1827.

14. J. B. Hirkude and A. S. Padalkar, *Appl. Energy*, 2012, **90**, 68.

15. C. Carraretto, A. Macor, A. Mirandola, A. Stoppato and S. Tonon, *Energy*, 2004, **29**, 2195.

16. H. Aydin and H. Bayindir, *Renewable Energy*, 2010, **35**, 588.

17. B. Lin, J. Huang and D. Huang, *Fuel*, 2009, **88**, 1779.

18. C. Oner and S. Altun, *Appl. Energy*, 2009, **86**, 2114.

19. C. Haşimoğlu, M. Ciniviz, İ. Özsert, Y. İçingür, A. Parlak and M. Sahir Salman, *Renewable Energy*, 2008, **33**, 1709.

20. M. Karabektas, *Renewable Energy*, 2009, **34**, 989.

21. X. Meng, G. Chen and Y. Wang, *Fuel Process. Technol.*, 2008, **89**, 851.

22. A. S. Ramadhas, C. Muraleedharan and S. Jayaraj, *Renewable Energy*, 2005, **30**, 1789.

23. S. Godiganur, C. H. S. Murthy and R. P. Reddy, *Renewable Energy*, 2010, **35**, 355.

24. S. Godiganur, C. H. S. Murthy and R. P. Reddy, *Renewable Energy*, 2009, **34**, 2172.

25. H. Raheman and S. V. Ghadge, *Fuel*, 2007, **86**, 2568.

26. M. Gumus and S. Kasifoglu, *Biomass Bioenergy*, 2010, **34**, 134.

27. N. Usta, E. Ozturk, O. Can, E. S. Conkur, S. Nas, A. H. Con, A. C. Can and M. Topcu, *Energy Convers. Manage.*, 2005, **46**, 741.

28. D. M. Korres, D. Karonis, E. Lois, M. B. Linck and A. K. Gupta, *Fuel*, 2008, **87**, 70.

29. A. Pal, A. Verma, S. S. Kachhwaha and S. Maji, *Renewable Energy*, 2010, **35**, 619.

30. M. A. Kalam and H. H. Masjuki, *Biomass Bioenergy*, 2008, **32**, 1116.

31. C. Lin and H. Lin, *Fuel Process. Technol.*, 2007, **88**, 35.

32. S. Jindal, B. P. Nandwana, N. S. Rathore and V. Vashistha, *Appl. Therm. Eng.*, 2010, **30**, 442.

33. G. Vijaya Raju Rao, G. Amba Prasad, P. Ramamohan, *Proceedings of the 16th National Conference on IC Engines and Combustion*, ed. P. K. Bose, Narosa Publishing House, New Delhi, 2000, p. 65.

34. N. R. Banapurmath, P. G. Tewari and R. S. Hosmath, *Proc. Inst. Mech. Eng., Part A.*, 2009, **223**, 31.

35. D. Sharma, S. L. Soni and J. Mathur, *Energy Sources, Part A*, 2009, **31**, 500.

36. A. Tsolakisa, A. Megaritis, M. L. Wyszynski and K. Theinnoi, *Energy*, 2007, **32**, 2072.

37. A. Keskin, M. Guru and D. Altíparmak, *Bioresour. Technol.*, 2008, **99**, 6434.

38. M. Guru, K. Atilla, C. Ozer, C. Can and S. Fatih, *Renewable Energy*, 2010, **35**, 637.
39. A. S. Ramadhas, S. Jayaraj and C. Muraleedharan, *Int. J. Global Energy*, 2008, **29**, 329.
40. T. C. Zannis, D. T. Hountalas and D. A. Kouremenos, SAE Paper No. 2004-01-0097SAE International, Warrendale, 2004.
41. A. Keskin, M. Guru and D. Altíparmak, *Fuel*, 2007, **86**, 1139.
42. G. R. Kannan, R. Karvembu and R. Anand, *Appl. Energy*, 2011, **88**, 3694.
43. M. A. Kalam and H. Masjuki, *Biomass Bioenergy*, 2002, **23**, 471.
44. E. Kinoshita, K. Hamasaki, C. Jaqin and K. Takasaki, SAE Paper No. 2004-01-1860, SAE International, Warrendale, 2004.
45. M. Iranmanesh, SAE Paper No. 2008-01-1805, SAE International, Warrendale, 2008.
46. D. H. Qi, H. Chen, L. M. Geng and Y. Z. H. Bian, *Renewable Energy*, 2011, **36**, 1252.
47. D. H. Qi, H. Chen, L. M. Geng, Y. Z. H. Bian and X. C. H. Ren, *Appl. Energy*, 2010, **87**, 1679.
48. S. Sinha and A. K. Agarwal, *J. Eng. Gas Turbines Power*, 2010, **132**, 042801.
49. A. K. Agarwal, *Proc. Inst. Mech. Eng.*, *Part D*, 2005, **219**, 703.
50. A. K. Agarwal, J. Bijwe and L. M. Das, *J. Eng. Gas Turbines Power*, 2003, **125**, 820.
51. R. Fraer, H. Dinh, P. Kenneth, L. R. McCormick, K. Chandle and B. Buchholz, SAE Report No. 2005-01-3641, SAE International, Warrendale, 2005.
52. R. K. Pandey, A. Rehman and R. M. Sarviya, *Renewable Sustainable Energy Rev.*, 2012, **16**, 1762.
53. A. N. Ozsezen, M. Canakci, A. Turkcan and C. Sayin, *Fuel*, 2009, **88**, 629.
54. F. Wu, J. Wang, W. Chen and S. Shuai, *Atmos. Environ.*, 2009, **43**, 1481.
55. Y. Ulusoy, R. Arslan and C. Kaplan, *Energy Sources*, *Part A*, 2009, **31**, 906.
56. C. Lin and R. Li, *Fuel Process. Technol.*, 2009, **90**, 883.
57. M. E. Tat, J. H. Van Gerpen and P. S. Wang, *Trans. ASABE*, 2007, **50**, 1123.
58. N. Y. Kado and P. A. Kuzmicky, Report No. SR-510-31463, National Renewable Energy Laboratory, Golden, Colorado, USA, 2003.
59. J. M. Luj, V. Bermez, B. Tormos and B. Pla, *Biomass Bioenergy*, 2009, **33**, 948.
60. S. Kalligeros, F. Zannikos, S. Stournas, E. Lois, G. Anastopoulos, C. Teas and F. Sakellaropoulos, *Biomass Bioenergy*, 2003, **24**, 141.
61. P. J. M. Frijters and R. S. G. Baert, *Int. J. Vehicle Des.*, 2006, **41**, 242.
62. K. Yoshiyuki, *JSAE Rev.*, 2000, **21**, 469.

63. S. Murillo, J. L. Míguez, J. Porteiro, E. Granada and J. C. Moran, *Fuel*, 2007, **86**, 1765.
64. J. Song and C. Zhang, *Proc. Inst. Mech. Eng.*, *Part D*, 2008, **222**, 2487.
65. H. Raheman and Phadatare, *Bio. Energy News*, 2003, 17.
66. B. Baiju, M. K. Naik and L. M. Das, *Renewable Energy*, 2009, **34**, 1616.
67. G. Knothe, C. A. Sharp and T. W. Ryan, *Energy Fuels*, 2006, **20**, 403.
68. M. S. Graboski, R. L. McCormick, T. L Alleman and A. M. Herring, Report No. SR-510-31461National Renewable Energy Laboratory, Golden, Colorado, USA, 2003.
69. A. Monyem, J. H. Van Gerpen and M. Canakci, *Trans. ASAE*, 2001, **44**, 35.
70. G. H. Abd-Alla, H. A. Soliman, O. A. Badr and M. F. Abd-Rabbo, *Energy Convers. Manage.*, 2001, **42**, 1033.
71. M. Lapuerta, J. M. Herreros, L. L. Lyons, R. García-Contreras and Y. Briceño, *Fuel*, 2008, **87**, 3161.
72. H. Kazunori, K. Eiji, T. Hiroshi, T. Koji and M. Daizo, presented at the 5th International Symposium on Diagnostics and Modeling of Combustion in Internal Combustion Engines, Nagoya, Japan, 2001.
73. D. Y. C. Leung, Y. Luo and T. L. Chan, *Energy Fuels*, 2006, **20**, 1015.
74. T. D. Durbin and J. M. Norbeck, *Environ. Sci. Technol.*, 2002, **36**, 1686.
75. C. S. Cheung, L. Zhu and Z. Huang, *Atmos. Environ.*, 2009, **43**, 4865.
76. M. A. R. Nascimento, E. S. Lora, P. S. P. Corr, R. V. Andrade, M. A. Rendon, O. J. Venturini and G. A. S. Ramirez, Conference paper, *Energy* (Oxford), 2008, **33**, 233. eds. C. A. Frangopoulos, C. D. Rakopoulos, G. Tsatsaronis, Aghia Pelagia, Crete, Greece, 2006.
77. D. H. Qi, L. M. Geng, H. Chen, Y. Z. H. Bian, J. Liu and X. C. H. Ren, *Renewable Energy*, 2009, **34**, 2706.
78. N. Usta, *Energy Convers. Manage.*, 2005, **46**, 2373.
79. C. Lin and H. Lin, *Fuel Process. Technol.*, 2007, **88**, 35.
80. P. Ye and A. L. Boehman, *Fuel*, 2012, **97**, 476.
81. O. Armas, K. Yehliu and A. L. Boehman, *Fuel*, 2010, **89**, 438.
82. M. Zheng, M. C. Mulenga, G. T. Reader, M. Wang, D. S. K. Ting and J. Tjong, *Fuel*, 2008, **87**, 714.
83. K. Ryu, *Bioresour. Technol.*, 2010, **101**, 78.
84. S. S. Gill, A. Tsolakis, J. M. Herreros and A. P. E. York, *Fuel*, 2012, **95**, 578.
85. X. Wang, C. S. Cheung, Y. Di and Z. Huang, *Fuel*, 2012, **94**, 317.
86. P. V. Bhale, N. V. Deshpande and S. B. Thombre, *Renewable Energy*, 2009, **34**, 794.
87. H. Kim and B. Choi, *Renewable Energy*, 2010, **35**, 157.
88. A. Gopinath, S. Puhan and G. Nagarajan, *Int. J. Energy Environ.*, 2010, **1**, 411.
89. M. Lapuerta, O. Armas and J. M. Herreros, *Fuel*, 2008, **1**, 25.
90. M. Alam, J. Song, R. Acharya, A. Boehman and K. Miller, SAE Paper No. 2004-01-3024, SAE International, Warrendale, 2004.

91. M. N. Nabi, M. S. Akhter and M. M. Z. Shahadat, *Bioresour. Technol.*, 2006, **97**, 372.
92. R. Ballesteros, J. J. Hernández, L. L. Lyons, B. Cabañas and A. Tapia, *Fuel*, 2008, **87**, 1835.
93. Y. Liu, T. Lin, Y. Wang and W. Ho, *J. Air Waste Manage.*, 2009, **59**, 163.
94. C. He, Y. Ge, J. Tan, K. You, X. Han and J. Wang, *Fuel*, 2010, **89**, 2040.
95. G. Fontaras, G. Karavalakis, M. Kousoulidou, T. Tzamkiozis, L. Ntziachristos, E. Bakeas, S. Stournas and Zissis Samaras, *Fuel*, 2009, **88**, 1608.
96. L. Turrio-Baldassarri, C. L. Battistelli, L. Conti, R. Crebelli, B. De Berardis, A. L. Iamiceli, M. Gambino, S. Iannaccone, *Sci. Total Environ.*, 2004, **327**, 147.
97. C. A. Sharp, S. A. Howell and J. Jobe, SAE Paper No. 2000-01-1968SAE International, Warrendale, 2000.
98. C. Peng, H. Yang, C. Lan and S. Chien, *Atmos. Environ.*, 2008, **42**, 906.
99. N. Arapaki, E. Bakeas, G. Karavalakis, E. Tzirakis, S. Stournas and F. Zannikos, SAE Paper No. 2007-01-0071, SAE International, Warrendale, 2007.
100. S. M. Correa and G. Arbilla, *Atmos. Environ.*, 2008, **42**, 769.
101. P. R. Costa Neto, L. F. S. Rossi, G. F. Zagonel and L. P. Ramos, *Quim. Nova*, 2000, **23**, 531.
102. Y. S. Zhang, J. Z. Yu, C. L. Mo and S. R. Zhou, SAE Paper No. 2008-32-0025, SAE International, Warrendale, 2008.
103. E. Zervas, X. Montagne and J. Lahaye, *Environ. Sci. Technol.*, 2002, **36**, 2414.
104. A. Munack, O. Schrder, J. Krahl and J. Bunger, *Agric. Eng. Int.*, *CIGR J. Sci. Res. Dev.*, 2001, **3**, EE-01-001.

Major Resources for Biodiesel Production

8.1 Introduction

Feedstocks such as soybean, palm oil and sunflower are considered to be first-generation biodiesel feedstocks because they were the first crops to be used to produce biodiesel and non-food feedstocks such as jatropha, mahua, jojoba oil, salmon oil, sea mango, waste cooking oils and animal fats are considered to second-generation feedstocks.[1] Researchers have also concentrated on third-generation biodiesel feedstocks which are derived from microalgae.[2] There are also fourth-generation biofuels that are genetically engineered to consume more CO_2 from the atmosphere than they will produce during combustion later as a fuel, and some fourth-generation technology include pyrolysis, gasification, upgrading, solar-to-fuel, genetic manipulation of organisms to secrete hydrocarbons, and conversion of vegetable oil and biodiesel into biogasoline using advanced technology.[3]

8.2 Non-Edible Oil Resources

8.2.1 *Argemone mexicana* L.

Argemone mexicana L. belongs to the family Papaveraceae also commonly known as Mexican prickly poppy because it is a species of poppy found in Mexico and now widely naturalized in the USA, India and Ethiopia.[4] The non-edible seed plant *A. mexicana* has been found to be very suitable for biodiesel purposes.[4] The plant grows in light (sandy) soils, requires well-drained soil and

Biodiesel: Production and Properties
Amit Sarin
© Amit Sarin 2012
Published by the Royal Society of Chemistry, www.rsc.org

can also grow in nutritionally poor soil. It can grow on wasteland and infertile land that is not used for agriculture. The seeds consist of 22–36% pale yellow non-edible oil, called argemone oil or katkar oil, which contains the toxic alkaloids sanguinarine and dihydrosanguinarine. The fatty acids present in the oil are myristic acid, palmitic acid, stearic acid, oleic acid, linoleic acid and arachidic acid.[5]

All parts of the plant, including the seeds contain toxic alkaloids, but the plant has many medicinal usages including as an analgesic, antispasmodic, possible hallucinogenic and sedative.[6,7]

8.2.2 *Azadirachta indica*

Azadirachta indica belongs to the family Meliaceae and its Indian common name is neem. The *A. indica* tree is a large evergreen tree usually 12–18 m high, present in Asia, Africa and Central and South America, which grows almost in all types of soil including clay, saline, alkaline, dry, stony and shallow soils and even in soils with a high calcareous content.[4] It also grows well in arid and semi-arid climates with maximum shade temperatures as high as 49 °C and rainfall is as low as 250 mm.[5] It reaches maximum productivity after 15 years and has a life-span of more than 50 years. A mature tree can produce 30–50 kg of fruit every year therefore there is a potential for the production of about 540 000 tons of seeds, which can yield about 107 000 tons of oil and 425 000 tons of cake.[4] Its oil is light to dark brown in color, bitter and has great potential as a feedstock for biodiesel to supplement other conventional sources. It has been reported that crude neem oil with a high free fatty acid content of 21.6% was pre-treated with an acid catalyst and the free fatty acid content was reduced to less than 1%. The reaction parameters for the two-step process were optimized so that the process gave a yield of 89% neem biodiesel[8] The major fatty acid composition is C16 : 0 (14.9%), C18 : 0 (14.4%) and C18 : 1 (61.9%).[9] Its iodine value is 69.3 and its cetane number (CN) 57.83. The oil is also used for preparing cosmetics, ayurvedic medicines and biopesticides.

8.2.3 *Calophyllum inophyllum*

Calophyllum inophyllum, commonly known as polanga in India and Penaga Laut in Malaysia, is a non-edible oilseed tree belonging to the Clusiaceae family.[10,11] The scientific name *Calophyllum* comes from the Greek word for 'beautiful leaf'. It is native to eastern Africa, southern coastal India, Southeast Asia, Australia and the South Pacific. It is a medium-to-large tree, normally up to 25 m tall, occasionally reaching up to 35 m and with diameter up to 150 cm. It has elliptical, shiny and tough leaves. The flower is about 25 mm and the fruit (ballnut) is a round, green drupe reaching 2–4 cm (0.8–1.6 inches) in diameter and having a single large seed.[11] When it is ripe, the fruit is wrinkled and its color varies from yellow to brownish-red. The grey, ligneous and rather soft nut contains a pale-yellow kernel which has a very high oil content (75%)

and the oil contains approximately 71% unsaturated fatty acids (oleic and linoloeic acids), therefore it has good low-temperature flow properties.[12] Fruit is usually borne twice a year and it produces up to 100 kg of fruit and about 18 kg of oil with 100–200 seeds per kg.[13] It grows in warm temperatures in wet or moderate conditions and requires an annual rainfall of around 1000–5000 mm.[11] Plantation can be at a density of 400 trees per hectare and the average oil yield is 11.7 kg oil per tree or 4680 kg oil per hectare.[5] It contains a large amount of protein and is non-edible due to the presence of saponins which are toxic. It has all the necessary properties for use as biodiesel.

Calophyllum inophyllum oil from the fruit traditionally has been used for medicine and cosmetics.[14,15] It has been used to treat wounds, facial neuralgia, skin ailments and hair loss, and rheumatism.

8.2.4 *Carnegiea gigantean*

Carnegiea gigantean commonly known as 'tall' is a conventional kraft pulp mill. Tall oil, a by-product of tall, is a mixture of triglyceride oils, fatty acids, resin acids, other terpenoids, and other related materials.[16] Fatty acids derived from tall oil mainly contains oleic acid, somewhat less linolenic acid and almost no linoleic acid.[17] Researchers have reported that there are opportunities to use tall oil components, as a source of raw material for biodiesel fuels, or as a source of additives for petrodiesel.[18]

8.2.5 *Cerbera odollam*

Cerbera odollam also known as sea mango belongs to the family Apocynaceae and the tree grows well in coastal salt swamps and creeks in south India and along riverbanks in southern and central Vietnam, Cambodia, Sri Lanka, Myanmar, Madagascar and Malaysia.[19] The tree grows to a height of 6–15 m and has dark green fleshy lanceolate leaves. The fruit, when still green, looks like a small mango with a green fibrous shell enclosing an ovoid kernel measuring approximately 2 × 1.5 cm and consisting of two cross-matching white fleshy halves. On exposure to air, the white kernel turns violet, then dark grey, and ultimately brown or black.[20] The sea mango tree is also known as the 'suicide tree' because of its highly poisonous nature and toxic content. The leaves and fruits of sea mango contain the potent cardiac substance (a glycoside) called cerberin, which is extremely poisonous if ingested.[21] The toxin can be easily separated out from the extracted oil by decantation and favors the potential of its oil for use as a biodiesel feedstock. The amount of oil extracted from *C. odollam* seeds is 54%. The fatty acid composition is mainly oleic (48.1%), followed by palmitic (30.3%), linoleic (17.8%) and stearic (3.8%).

Many parts of the tree are used for the manufacture of fibers. The latex is known in India for its emetic, purgative and irritant effects.[22]

8.2.6 *Croton megalocarpus*

The *Croton megalocarpus* plant belongs to the family Euphorbiaceae and is indigenous to East Africa being widespread in the mountains of Tanzania, Kenya and Uganda between the altitudes of 1300 and 2200 m in regions with annual rainfall between 800 and 1600 mm and average annual temperatures varying between 11 °C and 26 °C.[23] It is a dominant upper canopy forest tree reaching heights of 40 m or more. Croton seeds contain approximately 32% oil by weight.[24] The oil contains mainly C18 : 2 (72.7 wt%) and C18 : 1 (11.6 wt%). The biodiesel yielded has cloud and pour points of -4 °C and -9 °C, respectively, while its kinematic viscosity lies within the recommended standard value. This points to the viability of using croton biodiesel in cold regions.[23]

8.2.7 *Datura stramonium* L.

Datura stramonium L. belongs to the family Solanaceae.[25] It is native to Mexico and grows wild in warm and temperate regions in almost 100 countries, where it is typically found along roadsides and in dung heaps.[26] It grows in rich, light sandy soil or in calcareous loam and in open sunny areas, it can also grow in the shade and in a wide variety of climates (from humid to arid) and soil conditions, although it is very sensitive to frost.[25] It forms a bush with a height of approximately 0.3–1.5 m. It contains dangerous levels of deliriants and anticholinergic poisons such as atropine, hyoscyamine, and scopolamine.[27] The seeds contain 10.3–23.2 wt% vegetable oil.[28,29] A maximum fatty acid methyl ester yield (87%) and content of more than 98 wt% have been obtained.[25] It contains mainly linoleic acid (56.8 wt%) and oleic acids (28.5 wt%). It possesses the kinematic viscosity 4.33 mm^2 s^{-1} and the cold filter plugging point (CFPP) 5 °C.

D. *stramonium* was used for its medicinal potential by the Arab physician Avicenna in 11th Century Persia.[30] It has long been used as a herbal medicine with narcotic, anodyne, and antispasmodic effects.[31,32]

8.2.8 *Gossypium hirsutum* L.

Gossypium hirsutum L. belongs to the Malvaceae family and is commonly called cotton.[33] It is one of the second best potential sources for plant proteins after soybean and the ninth best oil-producing crop.[34,35] Cotton seed oil is extracted from the seeds of the cotton plant after the removal of cotton lint and contains significant amounts of saturated fatty acids, palmitic acid (22–26%), stearic acid in smaller amounts (2–5%) as well as traces of myristic, arachidic and behenic acids, lesser amounts of monounsaturated fatty acids, oleic acid (15–20%) being the major species, accompanied by traces of palmitoleic acid and a majority of diunsaturated linoleic acid (49–58%) with traces of linolenic acid.[33] The important fuel properties of cotton ethyl ester such as CN, kinematic viscosity, acid value, free and total glycerin compare well with ASTM D-6751 and EN 14214 specifications.[33]

8.2.9   *Hevea brasiliensis*

Hevea brasiliensis belongs to the family Euphorbiaceae and is the primary source of natural rubber. It is found mainly in Indonesia, Malaysia, Liberia, India, Sri Lanka, Sarawak, and Thailand.[4] The tree can grow up to 34 m in height and requires heavy rainfall. Rubber trees yield a three-seeded ellipsoidal capsule, each carpel with one seed and the seeds are ellipsoidal, variable in size, 2.5–3 cm long, mottled brown, lustrous, weighing 2–4 g each.[36] The oil content of the seeds, which may contain up to 17 wt% free fatty acids, ranges from 40 to 50 wt% and is high in unsaturated constituents such as linoleic (39.6 wt%), oleic (24.6 wt%), and linolenic (16.3 wt%) acids.[36] The physical properties of the resultant rubber seed oil methyl esters include cloud point and pour point values of 4 °C and −8 °C, respectively, and a kinematic viscosity (at 40 °C) of 5.81 mm^2 s^{-1}. The filtered oil is used as a feedstock for biodiesel production.

8.2.10   *Jatropha curcas* L.

Jatropha curcas L. is a large or medium shrub tree, up to 5–7 m tall, has a life-span of up to 50 years, and belongs to the Euphorbiaceae family, native to tropical America but widely distributed in tropical and subtropical regions throughout Africa, India and Southeast Asia.[37] It can grow on well-drained soils with good aeration and is a well adapted to marginal soil with low nutrient content, shedding its leaves in the dry season.[38,39] It can grow under a wide range of rainfall regimes from 250 to over 1200 mm per annum.[39,40] The plantation spaces of 2 × 2, 2.5 × 2.5 and 3 × 3 m are reported to give larger yields of fruit.[41] It bears fruits from the second year of its establishment, and the economic yield stabilizes from the 4th or 5th year onwards. The fruit is a kernel which contains three seeds each and it gives about 2–4 kg seeds per tree per year however in poor soils, the yields have been reported to be about 1 kg seeds per tree per year.[42] The oil yields are reported to be 1590 kg per hectare. Production of the seeds is about 0.8 kg m^{-2} per year and the oil content of seed ranges from 30 to 50% by weight and in the kernel from 45 to 60%. The fatty acid composition of jatropha classifies it as a linoleic or oleic acid type, which are unsaturated fatty acids therefore it has good low-temperature flow properties but not such good oxidation stability. The average production of jatropha oil on average soil is 1.6 million tons per hectare.

Fresh jatropha is a slow-drying, odorless and colorless oil and becomes yellow after aging. It is also known as Ratanjayot and physic nut. Jatropha oil is considered a non- edible oil due to the presence of toxic phorbol esters.[43,44] *Jatropha curcas* oil has been highlighted as a potential biodiesel feedstock among the non-edible oils. It satisfies all the biodiesel standards except the oxidation stability that is in the range of 2–4 h. Apart from being potential feedstock in the production of biodiesel as a diesel substitute, jatropha oil has other uses such as producing soap and biocides (insecticide, molluscicide, fungicide and nematicide).[45]

8.2.11 *Maduca indica*

Maduca indica commonly known as mahua or butter tree is a tropical tree belonging to the family Sapotaceae. It is approximately 20 m in height, found largely in the central and northern plains and forests of India, grows fast, possesses evergreen or semi-evergreen foliage, and is adapted to arid environments.[46,47] The medium-to-large deciduous tree is found abundantly in most parts of the world.[48] The mahua tree starts producing seeds after 4–7 years and continues for up to 60 years with the yield of *M. indica* seeds varying (5–200 kg per tree) depending upon the size and age of the tree.[49] The kernel constitutes about 70% of the seed and contains 50% oil.[50] The fatty acid composition of consists of both saturated (17.8–24.5% palmitic acid and 14–22.7% stearic acid) and unsaturated fatty acids (37.0–46.3% oleic acid and 14.3–17.9% linoleic acid).[51] The relatively high percentage of saturated fatty acids results in relatively poor low-temperature properties.

8.2.12 *Melia azedarach*

Melia azedarach belongs to the family Meliaceae and commonly known as syringa or Persian lilac. It is native to India, southern China, and Australia.[4] It is a deciduous tree that grows to 7–12 m in height. The oil content of dried syringa berries is around 10 wt% and it is characterized by a high percentage of unsaturated fatty acids such as oleic (21.8 wt%) and linoleic (64.1 wt%) acids.[52] Saturated fatty acids such as palmitic (10.1 wt%) and stearic (3.5 wt%) acids are also present. The physical properties of biodiesel prepared from syringa oil include a kinematic viscosity (at 40 °C) of 4.37 mm^2 s^{-1}, an iodine value of 127, and a specific gravity of 0.894 g mL^{-1}.

8.2.13 *Pongamia pinnata*

Pongamia pinnata commonly known as karanja, pongam, and honge belongs to the family Leguminaceae.[53] It is native to a number of countries including India, Malaysia, Indonesia, Taiwan, Bangladesh, Sri Lanka and Myanmar. *P. pinnata* is a fast-growing medium-size (12–15 m height) leguminous tree having the ability to grow on marginal land and this tree has been also successfully introduced to humid tropical regions of the world as well as parts of Australia, New Zealand, China, and the USA.[54,55] It requires an annual rainfall ranging between 500 and 2500 mm. Pongamia also exhibits wide adaptation to various soil types and saline/alkaline conditions, is regarded as a drought-tolerant species, can be planted in degraded lands, wastelands, or fallow lands and is highly tolerant to salinity. However, the highest growth rate and yield of oil are observed on well-drained soils with assured moisture.[56] *P. pinnata* bears flat-to-elliptical 5–7 cm long green pods which contain 1–2 kidney-shaped brownish red kernels and the yield of kernels per tree is between 8 and 24 kg, therefore, the tree has the potential for high oil seed production of about 25–40%.[57] Researchers have investigated the performance of CI diesel engines with Pongamia biodiesel and concluded that pongamia biodiesel can be used as an alternative fuel.[58] The predominant fatty acid is oleic acid (51.8%) with linoleic acid (17.7%), palmitic acid (10.2%), stearic acid (7.0%), and linolenic acid (3.6%) also present.[59]

It is currently cultivated mainly for ornamental purposes due to its large canopy and showy flowers.[4] Different parts of the plants such as the leaves, roots and flowers are reported to possess medicinal properties.

8.2.14 *Prunus sibirica* L.

Prunus sibirica L., a member of the family Rosaceae and the genus Prunus, is a deciduous shrub native to temperate, continental, mountainous regions, which include Eastern Siberia, the Maritime Territory of Russia, eastern and southeastern regions of Mongolia, and northern and north-eastern regions of China.[60] It is commonly known as Siberian apricot and can grow in temperate climates and thrives with abundant solar radiation, at low temperatures, in strong winds, low rainfall and poor soil. It bears fruit in its 4th year and enters into the full bearing period in its 8th year. The seed kernel has a high oil content (50.18 ± 3.92%), and the oil has a low acid value (0.46 mg g^{-1}) and low water content (0.17%). The fatty acid composition of the Siberian apricot seed kernel oil includes a high percentage of oleic acid (65.23 ± 4.97%) and linoleic acid (28.92 ± 4.62%). The measured fuel properties of the Siberian apricot biodiesel, except CN and oxidative stability, conform to EN 14214-08 and ASTM D-6751-10, in particular, the cold-flow properties were excellent.

The traditional use of Siberian apricot focuses on its ecological benefits, such as water and soil conservation, windbreak, sand fixation, environmental protection and greening.[61] The seed kernel oil can be used for edible oils, lubricants, cosmetics, surfactants, and in the prevention of cardiovascular diseases and the lowering of plasma cholesterol levels.[62]

8.2.15 *Putranjiva roxburghii*

The *Putranjiva roxburghii* tree belongs to the family Euphorbiacae and is also called the lucky bean tree. These plants are abundantly available in the Tropic of Cancer.[4] Putranjiva oil is yellow in color, highly pungent, volatile and rich in oleic acid. The composition of its oil is C14 : 0; 0.03%, C16 : 0; 10.23%, C16 : 1; 0.07%, C17 : 0; 0.07%, C17 : 1; 0.02%, C18 : 0; 10.63%, C18 : 1; 48.65%, C18 : 2; 27.50%, C18 : 3; 0.87%, C20 : 0; 1.05%, C20 : 1; 0.30%, C22 : 0; 0.24%, C22 : 1; 0.03%, and C24 : 0; 0.31%.[4] The viscosity of the oil is 37.6 cSt at 40 °C whereas at 100 °C the viscosity is 9.8 cSt and at higher temperatures the viscosity falls below 10 cSt which reduces atomization problems. The pure vegetable non-edible oil, can be used as an alternative diesel fuel without any modifications to the engine being required, in rural areas during fuel crises.[63]

8.2.16 *Ricinus communis* L.

Ricinus communis L. commonly known as castor is a drought-resistant plant believed to be originate from Abyssinia and distributed throughout tropical and subtropical regions. It is well adapted to temperate regions,[4] and belongs to the family Euphorbiaceae. The major castor-producing countries are India, China, Brazil, Russia and Thailand. The tree can reach up to 3 m height in temperate places but can reach 12 m high in tropical or subtropical climates. Castor oil has approximately 89.5% ricinoleic acid (also known as castor oil acid), an unsaturated fatty acid which is soluble in most organic solvents. Castor oil has properties such as unsaturated bonds, high viscosity, high molecular weight (298 g mol^{-1}), low melting point (5 °C), and a very low solidification point (-12 °C to -18 °C) making it industrially useful, low sulfur content (0.04%), ash content of 0.02%, negligible potassium, a heating value of 39.5 GJ t^{-1}, an iodine value of about 80 and a flash point of 260 °C.[4] The kinematic viscosity of transesterified castor oil is comparable to other vegetable oils making it suitable for use in biodiesel blends. It sequesters 34.6 tons of CO_2 per hectare and has good carbon trading potential. The genetic engineering of castor through silencing of the fatty hydroxylase gene, responsible for conversion of oleic acid to ricinoleic acid, leads to the accumulation of high levels of oleic acid which would be make it useful for the biodiesel market.[64,65]

As income generation from plantation of jatropha for the first 2–3 years is very low, castor can be inter-cropped to obtain income for that initial period as castor gives higher production yields in shorter periods than jatropha and this would help to improve the economical viability of both jatropha and castor plantations on a commercial scale.[66] Castor seeds and oils are used in the textile and printing industries, in the manufacturing of high-grade lubricants and also in traditional medicine.[67]

8.2.17 *Sapindus mukorossi*

The soapnut tree belongs to the family Sapindaceae and is generally found in tropical and subtropical climate areas in various parts of the world including Asia, the USA and Europe. The *S. mukorossi* species of soapnut grows wild from Afghanistan to China, ranging in altitudes from 200 to 1500 m in regions where precipitation varies from 150 to 200 cm per year.[68] The integration of soapnut tree plantations with community forestry in barren lands is considered to have great potential for the mass production of soapnut seeds, which are a potential non-edible oil feedstock for biodiesel production.[69] The major fatty acids in soapnut oil are oleic (9-Z-octadecenoic, C18 : 1) and eicosenoic (11-Z-eicosenoic, C20 : 1) acids, which contain one unsaturated C=C bond each.[70] Due to its fatty acid composition, soapnut oil biodiesel potentially exhibits superior oxidation stability.

The soapnut tree can be used for multiple applications such as rural building construction, oil and sugar presses, and agricultural implements among others. Soapnut has several applications in medicinal treatments, as a soap and surfactant and can be used for the treatment of soil contaminants.[71]

8.2.18   *Schleichera triguga*

Schleichera triguga commonly known as kusum belongs to the family Sapindaceae.[4] It is native to India, Myanmar, Sri Lanka, Timor and Java. It is a medium- or large-sized dense tree that can grow to 35 to 45 ft in height. The flowers come from February to April, it yields fruit in June and July and the fruits are smooth, hard-skinned berries containing one or two irregularly ellipsoidal slightly compressed seeds.[72] The kernel is u-shaped and the seed coat is brown. Most members of the Sapindaceae family contain cyanogenic compounds in the kernel and the presence of cyanolipids makes the oil toxic and non-edible. The cyanogenic compounds can be present in concentrations of 0.03–0.05% as hydrogen cyanide (HCN). Conventional saponification from kusum oil is not favored due to the presence of HCN in its effluents, therefore it is saponified only after pre-treatment.[72] The oil content is 51–62% but the yields are 25–36%. The performance of single cylinder diesel engine using kusum biodiesel has been reported.[73] However, as it contains HCN, biodiesel synthesized from kusum may posses HCN and which could be emitted on usage in the exhaust of CI engines thus, the exhaust emission coming from kusum oil biodiesel has to be analyzed and if it emits cyanide, its pre-treatment would be necessary.

The kusum oil has various uses in medicine, hair dressing, and for soap manufacture.[4] The kusum tree has economic importance as it can host the insects that produce a lacquer gum of high quality and therefore, it is also called by the name Lac tree.[72]

8.2.19   *Shorea robusta*

Shorea robusta commonly called the sal tree contributes nearly 5% of total forest area in India and its seeds yield 19–20% of a fatty oil (sal butter), which is contained in the cotyledons (kernels constitute 72% of the weight of the seeds).[74] It consists of a majority of stearic (44.2 wt%) and oleic (42.4 wt%) fatty acid methyl esters. Based on experiments, researchers concluded that the sal oil methyl ester can be used as a fuel without any modifications to the diesel engine and hence this biodiesel is a potential substitute for standard diesel fuel.[74]

Sal is the source of one of the most important commercial timbers widely used in constructional work where strength and durability are the main criteria, such as railway sleeper, beams, scantlings, floors, piles, bridges, carriage and wagon building, ladders, carts, spokes, hubs of wheels, tool handles, ploughs, dyeing vats, beer and oil casks.[74] Sal bark, which is available during the conversion of the logs, along with the leaves and twigs, is a promising tanning material. Sal dammar is used as incense and has also been successfully employed for hardening softer waxes for use in the manufacture of shoe polishes, carbon papers, typewriter ribbons *etc.* It is also used as an ingredient of ointments for skin diseases and for ear conditions. The de-oiled sal meal (sal oilcake) contains 10–12% protein and about 50% starch and is used as a cattle and poultry feed.

8.2.20 *Simmondsia chinensis*

Simmondsia chinensis belongs to the family Simmondsiaceae and is commonly known as jojoba (pronounced ho-ho-ba). It is a perennial shrub that is native to the Mojave and Sonoran deserts of Mexico, California, and Arizona.[4] It is also cultivated in Argentina, Israel, the USA and some Mediterranean and African lands.[75] It is drought-tolerant, grows in soils of marginal fertility, needs less water than many other crops, withstands salinity, apparently has a low fertilizer requirement and withstands desert heat without requiring much water or shade.[76] In the places where rainfall is less than 100 mm per year, jojoba plants persist as small stunted bushes and survive temperatures of up to 45 °C. Under more favorable conditions, jojoba plants thrive and potentially have a life-span of up to 100 years.[77,78] To date, there are no reports of jojoba plants being affected by catastrophic diseases or insect pests. It is 0.7–1.0 m high, possess thick, leathery, bluish green leaves and a dark-brown nutlike fruit. The jojoba seed has a nut-shaped form and is around 1–2 cm long, with a red-brown to dark-brown color and the seed contain between 45 and 55 wt% of jojoba oil–wax.[79] Jojoba oil is not a triglyceride but a mixture of long-chain esters (97–98 wt%) of fatty acids and fatty alcohols, and therefore is more properly referred to as a wax; however, Jojoba oil–wax is the term in general use. It contains minor amounts of free fatty acids and alcohols, phytosterols, tocopherols, phospholipids and trace amounts of a triacylglycerol.[80] The fatty acid component of jojoba wax esters primarily consists of eiconenoic, erucic, and oleic acids with *cis*-11-eicosen-1-ol and *cis*-13-docosen-1-ol principally comprising the alcohol component.

One of the first uses of the jojoba oil–wax in the 1970s was the substitution of sperm whale oil, when whale hunting was banned worldwide, but it has other uses in cosmetics, pharmaceuticals, dietetic foods, animal feeding, lubrication, polishing and gardening.[81]

8.2.21 *Sterculia striata*

The *Sterculia striata* tree belongs to the Sterculiacea family and is also commonly known as chicha.[82] It has its origin in India and Malaysia and is also common in northeast Brazil and can adapt to semi-arid lands. The plant starts to produce fruit within 18–24 months after plantation and an adult tree with adequate agricultural practices can produce about 40 kg of seeds per year. The seeds contain up to 60% kernels that contain oil at concentrations varying from 28 to 32%.[83] The fatty acid composition shows that the oil is rich in fatty acids containing a cyclopropene ring in their chain, mainly sterculic (octadec-9,10-methylene-9-ene) and malvalic (heptadec-8,9-methylene-8-1-ol) acids.[84] This tree can be explored further for the production of biodiesel.

8.2.22   *Terminalia belerica Roxb.*

Terminalia belerica Roxb. commonly known as bahera belongs to the Combretaceae family.[85] It is found in deciduous forests throughout the greater part of India, particularly in the Indian peninsula in moist valleys. It can tolerate moderate drought and heavy rainfall and it is a light-demander, hardy, sensitive to frost, and a good coppicer. It can be cultivated in almost all types of soil, however, it prefers loamy soil, red soil, and the silted soil of tropical and subtropical areas. The tree can attain up to 30–40 m height and 1.8–3.0 m in girth. It matures in 6–8 years and yields about 500 kg of raw fruit annually. The fruit is globular, obscurely five-angled, 1 inch long, grey and velvety, pulpy and rather dry and contains a hard thick-shelled stone. The fruit ripens at the end of rainy season and is collected during winter.[1] The seed germinates within 14–30 days.

The oil yield of the *T. belerica* seed is about 31% (dry weight) and the fatty acid profile shows the predominance of oleic acid (C18 : 1; 61.5%) along with linoleic (18.5%) and palmitic (11.6%) glycerides.[85] Its biodiesel properties such as viscosity, CN, density and others key properties are within the ASTM D-6751 biodiesel standard. An experimental study revealed the possibility of *T. belerica* seed oil as potential source of biodiesel.

T. belerica is also reported to have therapeutic value for the treatment of liver and digestive disorders and the dry fruits also possesses potential broad-spectrum antimicrobial activity.[86,87] It is also routinely used as a traditional medicine to treat several ailments, such as fever, cough, diarrhea, skin diseases and oral thrush.[85]

8.2.23   *Thespesia populnea* L.

Thespesia populnea is a member of the family Malvaceae, commonly known as the portia tree.[88] It is a native of Hawaii, may have been introduced by the early Polynesians and is now widely distributed in coastal areas of the tropics and subtropics.[89] It is a fast-growing small tree ranging in height from 6 to 10 m. The seeds of *T. populnea* contain about 20% (w/w) oil. *T. populnea* is the first biodiesel fuel to be produced from a feedstock containing cyclic fatty acids and contains 8,9-methylene-8-heptadecenoic (malvalic) and smaller amounts of two cyclopropane fatty acids besides greater amounts of linoleic, oleic and palmitic acids.[88] The total unsaturated fatty acids in *T. populnea* and its methyl ester amounts to approximately 58% and the total saturated fatty acids to approximately 32–33% with the remaining percentage being accounted for by the fatty acids with cyclic moieties as well as other unidentified species.

The density, viscosity, CN and higher heating value of biodiesel from *T. populnea* are competitive with biodiesel fuels derived from other feedstocks.[88] The presence of cyclic fatty acids did not have any significant effect on its fuel properties. The cloud, pour and CFPPs were relatively high. The ash content, acid value and copper strip corrosion value are also within the limits prescribed by ASTM.

T. populnea seed oil is used in the treatment of cutaneous infections such as scabies, psoriasis, ringworm, guineaworm, eczema, and herpetic diseases in India.[90,91] The bark, roots and fruits of have also been employed in the folk medicine system for curing dysentery, cholera, wounds, and hemorrhoids.[91] The plant is used for producing rope and dye, and is valued as a shade tree and a windbreaker to control soil erosion.

8.2.24 *Thevettia peruviana*

Thevettia peruviana, commonly known as yellow oleander, belongs to the family Apocynaceae. It is native to tropical South America especially Mexico and Brazil and is also native to the West Indies and other tropical regions of the globe.[4] It is a perennial shrub which can grow to a height of 4.5–6 m. It has deep-green linear sword-shaped leaves and funnel-shaped (yellow, white or pinkish-yellow colored) flowers and the plant starts flowering after one and a half years. After that it blooms thrice a year and produces more than 400–800 fruits yearly depending on the rainfall and plant age.[92,93] It has a high oil content (67%) in its kernel.[5] The shell contains some hexane-soluble material (0.36%) therefore the shell should be separated before oil expulsion.[4] The plant has an annual seed yield of 52.5 tons per hectare and about 1750 L of oil can be obtained from a hectare of wasteland. Its kernel oil has very good thermal stability.[94] Its major fatty acid composition is C16 : 0 (25.4%), C18 : 1 (46.8%) and C18 : 2 (27.8%). Its iodine value is 92.6 and its CN is 52.22.

8.2.25 *Vernicia montana*

There are two major species of tung oil trees: *Vernicia montana* and *Vernicia fordii* (Hemsl.) belonging to the family Euphorbiaceae.[95] The *V. Montana* tree grows up to 20 m in height and is usually dioecious with brown bark, glabrous branches, and sparsely elevated lenticels. It is found throughout open forests below an altitude of 1600 m in China, Myanmar, Taiwan, Thailand and Vietnam, and it is cultivated in Japan and Malawi.[96] The compressed globose seeds have a thick verrucose seed coat. The oil content of the seeds is approximately 21 wt%, while the content in whole nuts is 41 wt%; oil is produced in 300 to 450 kg per hectare amounts. The oil is also called 'abrasin oil' or 'Chinese wood oil'. The major fatty acid in tung oil is $9Z,11E,13E$-α-elaeostearic acid, which contains three conjugated double bonds (C18 : 3).[97] The biodiesel produced from tung oil has a CFPP of −11 °C, but has a poor oxidation stability of 0.3 h because of the large amount of α-elaeostearic acids.[98,99] *V. montana* is considered a potential non-edible oil feedstock for biodiesel production. Tung oil is also a source of drying oils used in paints, in varnishes and for polymerization.[100]

8.2.26 *Ziziphus mauritiana*

Ziziphus mauritiana belonging to the family Rhamnaceae is a fast-growing small tree.[5] It can grow in a temperature range of 13– 48 °C. It is drought-resistant and frost-hardy, grows best in sandy loam soil and wastelands of arid and semi-arid regions and can be utilized for its plantation.[5,101] It can grow at altitudes of 1500 m with rainfall of 150–2250 mm. It can be sown by seeding, seedling or grafting and matures after five years with a life-span of approximately 25 years. The oil yield from the seeds is 33%. The reported iodine value is 81.8 and the CN is 55.37. Its fatty acid profile is C16 : 0(10.4%), C18 : 0(5.5%), C18 : 1(64.4%), C18 : 2(12.4%), C20 : 0(1.8%), C20 : 1(2.6%), C22 : 0(1.2%) and C22 : 1(1.7%).[5,101] *Ziziphus mauritiana* could be potential biodiesel feedstock if explored properly.

8.3 Edible Oil Resources

8.3.1 *Arachis hypogea* L.

Arachis hypogea L. commonly known as peanut or groundnut belongs to the Fabaceae (legume, pea or bean) family and is grown in China, India, the USA and elsewhere.[102] Peanut oil comprises 40–50% of the mass of dried nuts. The National Agricultural Statistics Service of the USDA reported that the six-year (2005–2010) average yield of peanuts was 3563 kg per hectare per year which results in a calculated yield for peanut oil of 1425–1782 kg per hectare per year.[103] The potential importance of peanut oil as a fuel was cited by Mr Rudolf Diesel. The percentage of oleic acid in traditional peanut oil ranges from 41 to 67%, whereas high-oleic cultivars contain close to 80% of this constituent.[104] Biodiesel prepared from peanut oil has been reported in the literature.[105]

8.3.2 *Asclepias syriaca*

Asclepias syriaca commonly known as milkweed is native to the northeast and north central USA where it grows on roadsides and in undisturbed habitat.[106] The seed contains 20–25 wt% oil that is currently underutilized and may provide an inexpensive source of triglycerides for conversion to biodiesel fuel. Milkweed seed oil is composed of over 90% unsaturated fatty acids with nearly

50% linoleic acid and less than 2% linolenic acid. Based on its fatty acid profile, the oil is expected to provide an alternative source of biodiesel fuel. The kinematic viscosity, oxidative stability, and lubricity values compare favorably to soybean esters.

Milkweed seed is commercially harvested for the seed floss which exhibits a high insulating value and has applications as a fiber fill material in hypoallergenic comforters and pillows and the seed floss may also be combined with cotton fibers for woven textiles.[107] Historically, the plant is a source of fiber and medicinals.[108] The milky sap after which it is named has also been examined as a source of latex.[109]

8.3.3 *Balanites aegyptiaca*

Balanites aegyptiaca, commonly known as desert date, belongs to the family Zygophyllaceae. If proper cultivation practices are used, with emphasis on low-quality irrigation water, these trees can be extremely well developed in hyper-arid conditions.[110] Each tree can yield up to 52 kg date fruits, and the kernels oil content may reach up to 46.7% (based on dry weight). A good-condition mature desert date tree produces as many as 10 000 yellow date-like fruits annually. Each fruit, weighing 5–8 g, consists of an epicarp (5–9%), a mesocarp or pulp (28–33%), an endocarp (49–54%) and a kernel (8–12%). The oil content of desert date kernels approaches approximately 50%. The oil consists of four major fatty acids: palmitic (16 : 0), stearic (18 : 0), oleic (18 : 1), and linoleic (18 : 2), constituting 98–100% of the total fatty acids in the oil in all tested genotypes.[110] Linoleic acid was the most prevalent fatty acid, ranging from 31 to 51% of the fatty acid profile. It is highly adapted to the drier parts of Africa and south Asia and distributed in the most adverse arid desert environments of the world.[111]

8.3.4 *Brassica carinata*

Brassica carinata, a native plant of the Ethiopian highlands is widely used as food by the Ethiopians, therefore it is commonly also known as Ethiopian mustard oil.[112] It belongs to the family Brassicaceae. The agronomic performance and the energetic balance described in areas such as Spain, California and Italy have confirmed that *B. carinata* adapts better and is more productive both in adverse conditions (clay- and sandy-type soils and in semi-arid temperate climates) and under low-cropping system. The acid profile by weight % includes palmitic acid (C16) 3.1, stearic acid (C18) 1.0, oleic acid (C18 : 1) 9.7, linoleic acid (C18 : 2) 16.8, linolenic acid (C18 : 3) 16.6, arachidic acid (C20) 0.7 and erucic acid (C22 : 1) 42.5. The biodiesel, produced by transesterification of the oil, displays physicochemical properties suitable for its use as a diesel car fuel.[112]

8.3.5   *Brassica napus* L.

Brassica napus L. commonly known as rapeseed or canola is a cruciferous crop and is the most significant raw material for the biodiesel-producing industry in Europe and Canada.[17] the seeds are small and hot, dry conditions can limit their oil content. Canola seed has more than 43% oil content, higher than many other oilseeds including soybeans, which contain approximately 18% oil, and the new quality standards set for variety registration in western Canada will see the oil content in new varieties increase.[113] This change will ensure canola continues to deliver more oil per unit seed, improving processing efficiency. Canola methyl ester has approximately 7% saturated fat and the remaining 93% is unsaturated. Out of the 93%, oleic acid dominates (62.4%), then linoleic acid (19.73%) and linolenic acid (9.47 %).[114] The low saturated fat content of canola oil means improved cold-weather performance of the biodiesel. Rapeseed biodiesel has viscosity 4.439 cSt at 40 °C, CN 54.40, cloud point −3.3 °C, CFPP −13 °C and oxidation stability 7.6 h.[17] It also contains 10% oxygen by weight that leads to a reduction in emissions of hydrocarbons, toxic compounds, carbon monoxide and particulate matter.[113] However, there is concern about the use of rapeseed oil for biodiesel production because rapeseed is presently grown with a high level of nitrogen-containing fertilizer which generates N_2O, a potent greenhouse gas with 296 times the global warming potential of CO_2.[115]

8.3.6   *Cocus nucifera*

Cocus nucifera commonly known as coconut is widely harvested in tropical coastal areas and its oil accounts for nearly 20% of all the vegetable oil produced in the world.[17] Coconut oil is extracted from the copra, the dried flesh of the nut and it remains solid at a relatively higher temperature than most other vegetable oils. The oil content is approximately 65%. The main disadvantage of using coconut oil in engines is that it starts solidifying at temperatures below 22 °C and by 14 °C it does not flow at all. The fatty acid composition includes myristic acid (13–19%), linoleic acid (0–1%), oleic acid (5–8%) and palmitic acid (8–11%) with others in trace amounts.[17]

8.3.7   *Corylus avellana*

Corylus avellana or hazelnut or filbert belongs to the family Betulaceae and is an important commercial crop in the USA and elsewhere.[116] It contains 51–75 wt% plant oil, which equates to a production potential of approximately 1000 kg oil per hectare per year.[117] The primary fatty acid constituents are oleic (68–78%), linoleic (14–23%) and palmitic (4–6%) acids. Preparation of biodiesel from hazelnut oil has been reported.[118]

8.3.8 *Cucumis melo*

Cucumis melo commonly known as muskmelon belongs to the Cucurbitaceae family.[118] It originates from Iran and Pakistan and is mostly grown in the warmer regions of the world.[119,120] In addition to a good source of protein, muskmelon seeds contain vegetable oil varying from 35–49%.[120,121] Its seeds are classified as a waste product. The seed oil contains linoleic acid in the range 50– 69% followed by oleic (16.8–21%), palmitic (8.4–17.68%) and stearic (4.6–10.68%) acids.[118,122] The fuel properties of its methyl ester found to satisfy the ASTM D-6751 and EU 14214 specifications.[118]

Muskmelon seeds are also used in medicines for diabetes and chronic or acute eczema.[122]

8.3.9 *Elaeis guineensis*

Elaeis guineensis is a variety of the popularly known palm tree which is native to West Africa where it first grew wild and was later developed into an agricultural crop.[123] It is a tropical perennial plant and grows well in lowland with humid places and therefore it can be cultivated easily in Malaysia, Indonesia and Thailand.[124] The world production of palm oil is 45 million tons and the highest production is in Southeast Asia with a total 89% of total palm oil production (40% in Malaysia, 46% in Indonesia, 3% in Thailand). It can produce 10–35 tons per hectare of fresh fruit bunch annually.[125] The tree can grow up to 20–30 m in height.[126] The fruitlet of palm consists of a fibrous mesoscarp layer and the shell containing the kernel which contains oil and carbohydrate reserves for the embryo.[127,128] The shell contains about 49% palm oil and the kernel about 50% palm kernel oil. It is reported that the economic life of an oil palm tree is 20–25 years of its life-span of 200 years. The yield of crop is 4–5 tons of oil per hectare annually. The fatty acid profile in weight % consists of palmitic (C16 : 0) 40.3, stearic (C18 : 0) 4.1, oleic (C18 : 1) 43.4 and linoleic (C18 : 2) 12.2 acids. The average yield of approximately 6000 L of palm oil per hectare can produce 4800 L of biodiesel.[129]

8.3.10 *Glycine max*

Glycine max is commonly known as soybean and with a production of about 222 million tons, it is the most important oil-bearing plant cultivated worldwide, particularly in the USA, Brazil and Argentina.[130] Soybeans can be produced without or with nearly zero nitrogen. It is reported that 5546 kg of soybeans are required to produce 1000 kg of oil, and biodiesel production using soybean requires 27% more fossil energy than the biodiesel fuel produced.[131] The fatty acid composition of soybean biodiesel includes linoleic acid (53.27%), oleic acid (23.12%), palmitic acid (10.95%) and linolenic acid (6.77%) with others in trace amounts.[114] The reported soybean biodiesel viscosity is 4.039 cSt at 40 °C, its CN is 45.00, cloud point 1.0 °C, CFPP −4 °C and oxidation stability 2.1 h.[17]

8.3.11 *Guindilia trinervis*

Guindilia trinervis belongs to the family Sapindaceae, its common name is guindilla, and it is native to Chile and Argentina.[132] The plants can survive in soils and climates regarded unsuitable for agriculture. *Guindilia trinervis* is a small shrub that grows abundantly at altitudes over 1500 m, and is usually covered by snow during the winter. Each adult plant occupies an area of about 1.5 × 1.5 m and has simple, sessile leaves and a three-part fruit (1–1.5 cm in diameter) in which one or two carpels sometimes abort to make the fruit apparently one-seeded.[133] The seed has two cotyledons wrapped inside a seedcoat and contained in a round, thick hull 1–1.5 cm in diameter. The oil yield of whole seeds is about 30% w/w and under natural conditions it may be possible to obtain 3.5–4.0 tons of whole seeds per hectare or about 1000 L oil per hectare.[132] Its main fatty acids RE 62.3% oleic, 12.9% gadoleic, 10.1% linoleic and 9.6% palmitic.[134,135] Transesterification reactions yielded a biodiesel with ester content >99%; CN 59; oxidative stability at 110 °C, 18.9 h; kinematic viscosity at 40 °C, 4.867 mm^2 s^{-1}; CFPP, 4 °C; sulfur content 1.0 mg kg^{-1}; sulfated ash < 0.01% p/p; acid value 0.024 mg KOH per g and phosphorous content (<0.5 mg kg^{-1}).[132] All values are within European and USA specifications.

8.3.12 *Helianthus annuus*

Helianthus annuus belongs to the family Asteraceae and is commonly known as sunflower. It is one of the more prominent oilseed crops for biodiesel production and it can grow in a variety of climatic conditions but is considered to be an inefficient user of nutrients.[17] The average yield is 952 L per hectare and oil content is 25–35%.[132] The fatty acid composition is C16 (3.0–6.0%), C18 (1.0–3.0%), C18 : 1 (14.0–35.0%), C18 : 2 (44.0–75.0%), C18 : 3 (<1.5%) and other trace acids.[136] The reported viscosity is 4.439 cSt at 40 °C, the CN is 54.4, the cloud point is 3.4 °C, the CFPP –3 °C and the oxidation stability is 0.9 h.[17]

8.3.13 *Juglans regia* L.

Juglans regia L. belongs to the Juglandaceae family and is popularly known as walnut.[116] It is indigenous to southeastern Europe, eastern Asia and North America.[137] Worldwide, walnuts rank second (1.4 million metric tons; MMT) behind almonds and ahead of hazelnuts (0.73 MMT) in tree/shrub nut production and China and the USA accounted for 75% of global commercial walnut production in 2010, with China producing a greater amount due to having a larger area under cultivation.[138] Walnut kernels contain around 50–70% oil by weight, depending on the cultivation, geographic location and availability of water.[139] The major fatty acid constituents reported in walnut oil are linoleic (57–62%), oleic (12–20%), linolenic (11–16%) and palmitic (6–8%) acids. To date, there are no reports on the preparation or fuel properties of biodiesel from walnut oil.[116]

8.3.14 *Moringa oleifera*

Moringa oleifera belongs to the family Moringaceae.[140] It is indigenous to sub-Himalayan regions of northwest India, Africa, Arabia, Southeast Asia, the Pacific and Caribbean Islands and South America, and is now distributed in the Philippines, Cambodia and Central and North America.[141] It thrives best in a tropical insular climate and is plentiful near the sandy beds of rivers and streams, however it is also drought-tolerant and can tolerate poor soil, a wide rainfall range (25 to >300 cm per year), and soil pH from 5.0 to 9.0.[142,143] It can grow to heights of 5–10 m, and its dried seeds are round or triangular shaped, and the kernel is surrounded by a lightly wooded shell with three papery wings.[143,144] *M. oleifera* seeds contain 33 – 41% w/w of vegetable oil.[5] The oil contains more than 70% oleic acid.[140,144]

The oil also contains a disproportionately high content (7.1%) of behenic (docosanoic; 22 : 0) acid.[140] *M. oleifera* is commonly known as 'ben oil' or 'behen oil', due to its content of behenic (docosanoic) acid, and it possesses significant resistance to oxidative degradation.[145] As the oil has a high content of oleic acid with saturated fatty acids comprising most of the remaining fatty acid profile, the methyl esters obtained from this oil exhibit a high CN of approximately 67, one of the highest found for a biodiesel fuel.[140] *M. oleifera* oil could be an acceptable feedstock for biodiesel.

M. oleifera has many medicinal uses and has significant nutritional value.[146]

8.3.15 *Oryza sativum*

Oryza sativum is commonly known as rice bran and belongs to the family Poaceae. It is a co-product of rice milling and it contains 15–25% oil which can be used as a vegetable oil for transesterification with alcohol to produce methyl esters.[147] It is available in large quantities in rice-cultivating countries.[148] The estimated potential yield of crude rice bran oil would have been about 8 million metric tons, if all rice bran produced in the world in 2005 had been harnessed for oil extraction.[149] Physicochemical characterization indicated that the kinematic viscosity of rice bran oil methyl ester is near to that of diesel fuel and its CN is also higher than diesel fuel.[150]

8.3.16 *Pistacia chinensis*

Researchers have considered *Pistacia chinensis* (common name Chinese Pistache and from the family Anacardiaceae) oil as an alternative for producing biodiesel in China.[151] It is a small-to-medium-sized tree native to central and western China and it is hardy, can withstand harsh conditions and poor quality soils, and grows to 9–15 m in height, exceptionally up to 25 m.[152] It has been reported that the oil content in the seeds is more than 40%, while that in the kernels is 56.7%; and the oil is not toxic, but edible.[153]

An experimental study of the performance and emissions of a diesel engine using two different biodiesels derived from Chinese pistache oil and jatropha oil compared with pure diesel has been conducted.[151] The results show that the diesel engine works well and the power outputs are stable running with the two selected biodiesels at different loads and speeds.

8.3.17   *Sesamum indicum* L.

Sesamum indicum L. is an oilseed herbaceous crop of the Pedaliaceae family and is commonly known as sesame.[154] It is thought to have originated from Africa and Turkey. It is also widely cultivated in many parts of the world, primarily in tropical and subtropical areas, including India, China, Sudan, Burma, Tunisia, Egypt, Thailand, Mexico, Guatemala, El Salvador, Afghanistan, Pakistan, Bangladesh, Indonesia, Sri Lanka, Saudi Arabia and Turkey, and has recently been adapted to semi-arid regions.[155,156] It contains oil (44–58%), protein (18–25%), carbohydrate (13.5%) and ash (5%). The most abundant fatty acids are oleic (43%), linoleic (35%), palmitic (11%) and stearic (7%) acids, which together comprise about 96% of the total fatty acids.[155] The experimental results support that the methyl ester of sesame seed oil can be successfully used as biodiesel.[154]

Sesame is used in food, nutraceuticals, pharmaceuticals and industry in many countries because of its high oil, protein and antioxidant contents.[156] The fat of sesame is of importance in the food industry due to its flavor and stability, and because it can be used to cook meals of high quality.

8.3.18   *Syagrus coronata* (Mart.) Becc.

Syagrus coronate belongs to the family Arecaceae and is popularly known in Brazil as licuri or ouricuri.[157] The species grows mainly in semi-arid regions and, in Brazil, its geographical distribution comprises part of the Minas Gerais, Bahia, Sergipe and Alagoas states. The *S. coronate* tree bears fruit throughout the year but, however, March, June and July are the most favorable months for production. The oil is obtained from the fruit, with yields of 39% by mass. The main component is lauric acid (12 : 0) 42.0%.?The properties of the biodiesel produced are comparable with those commonly accepted for use as biodiesel.

8.3.19   *Xanthoceras sorbifolia* Bunge.

Xanthoceras sorbifolia Bunge. belongs to the family Sapindaceae, is commonly known as yellow horn and has proved to be an ideal energy crop, which can be widely planted against drought, cold, salt and starvation.[158] It is native to northern China. It has a high oil content (55–65%). It mainly contains linoleic acid methyl ester (42.12%), oleic acid methyl ester (30.4%) and eicosenoic acid methyl ester (10.09%) with other trace fatty acids. The viscosity is reported to be 4.4 cSt at 40 °C and its CN 56.1. Due to its high unsaturated fatty acid content (85–93%) and low acid value (0.5–0.7 mg KOH per g), it can be used as a high-quality feedstock for biodiesel production.

8.4 Other Resources

8.4.1 Algae/Microalgae

Microalgae are prokaryotic or eukaryotic photosynthetic microorganisms that can grow rapidly and live in harsh conditions due to their unicellular or simple multicellular structure. They are categorized into four main classes: diatoms, green algae, blue–green algae and golden algae.[159,160] Examples of prokaryotic microorganisms are cyanobacteria (*Cyanophyceae*) and eukaryotic microalgae are for example green algae (*Chlorophyta*) and diatoms (*Bacillariophyta*).

Researchers have reported that algae cultivation has four basic, and equally important, requirements; carbon, water, light and space.[161] CO_2 needs to be provided at very high levels, much higher than can be attained under natural conditions and the flue gases from industrial processes, and in particular from power plants, are rich in CO_2 that would normally be released directly into the atmosphere and thereby contribute to global warming. Researchers suggested that by diverting the CO_2 fraction of the flue gas through an alga cultivation facility, the CO_2 could be diverted back into the energy stream and the rate of algal production can be greatly increased. Water, is the second requirement and high-nutrient wastewater from domestic or industrial sources, which may already contain nitrogen and phosphate salts, can be added to the algal growth media directly and this allows for algae production to be improved cheaply, while simultaneously treating wastewater. Salt water can be used, either from saline aquifer or sea water. Abundant light is often accomplished by situating the facility in a geographic location with abundant, uninterrupted sunshine and when working with bioreactors, sunlight quantity and quality can be further enhanced through the use of solar collectors, solar concentrators, and fiber optics in a system called a photobioreactor. By maximizing the quality and quantity of these requirements, it is possible to maximize the quantity of oil-rich biomass and the return on investment.

There is a growing interest in algae-based biodiesel for its higher yield of non-edible oil, and because it does not compete for land with food production.[1] Algae are considered as the most promising non-food source of biofuels as they have a rapid reproduction rate and can grow in salt water, harsh conditions, submerged areas and sea water and it is estimated that more than 50 000 species exist, but only a limited number, of around 30 000, have been studied and analyzed.[162]

Algae can accumulate very high levels of lipid that can be easily transesterified to biodiesel; the oil content in microalgae can exceed 80% by weight of dry biomass.[163] Biodiesel produced from microalgae is not resource-limited and has the potential for yields 50–100 times greater than those of biodiesel from crops. The biofuel from algae contains no sulfur, is non-toxic and highly biodegradable. It is estimated that 1 kg of dry algae biomass utilizes about 1.83 kg of CO_2, therefore microalgae biomass production can help in the bio-fixation of waste CO_2 with respect to air quality maintenance and improvement.[164]

It was found that algae-based biodiesel has a superior yield per hectare over conventional oil crops because algae can be grown in a farm or bioreactor, but the main obstacle for the commercialization of algae-based biodiesel is its high production cost from requiring high-oil-yielding algae strains and effective large-scale bioreactors.[165,166] Biodiesel produced from microalgae has been found to have properties such as density, viscosity, flash point and other parameters which comply with the limits established by the ASTM for biodiesel quality.[2] Microalgae lipids are mostly neutral lipids due to their high degree of saturation, and their accumulation in the microalgal cell at different stages of growth (depending on the strain) makes microalgal lipids a potential diesel fuel substitute.[167,168]

Microalgae produce valuable co-products or by-products such as biopolymers, proteins, carbohydrates and residual biomass, which may be used as feed or fertilizer. In addition, cultivation of microalgae does not require herbicides or pesticides.[169] Microalgae are considered to be an efficient biological system for harvesting solar energy to use in the production of organic compounds and because of their small size, they can be easily chemically treated.[170]

8.4.2   Animal Fats

The fats derived from animals can be another group of feedstock for biodiesel production, for example the animal fats used to produce biodiesel include lard and tallow.[171,172] Animal fats are often priced favorably for conversion into biodiesel.[173] Animal fat methyl ester has a high CN but is low in free fatty acids and water.[174,175]

8.4.3   Municipal Sewage Sludges

Sewage sludge is an abundant organic waste or by-product generated in wastewater treatment plant (WWTP) facilities after primary and secondary treatment processes.[176,177] WWTP facilities produce huge amounts of sludge per year, for example, WWTP facilities in the USA alone produce over 6.2 million metric tons of dry sludge every year and six wastewater treatment plants in London, Canada produced 3.8×10^5 m^3 wastewater sludge in 2008.[178,179] Research has indicated that the lipids contained in sewage sludge are a potential feedstock for biodiesel and it contains a significant amount of lipid.[178] Lipid is a natural mixture of triglycerides, diglycerides, monoglycerides, cholesterols, free fatty acids, phospholipids, sphingo-lipids *etc.*[180] Lipids are usually extracted from the sludge with organic solvents to avoid interference in the biodiesel synthesis, however, the lipid extraction for biodiesel production is a difficult process.

8.4.4 Waste Cooking Oil

Waste cooking oil (WCO) is an oil-based substance which consists of vegetable matter that has been used in cooking or preparation of foods and is no longer suitable for human consumption.[181] WCO is categorized by its free fatty acid content, for example, if the content of WCO is <15%, then it is called 'yellow grease'; otherwise, it is called 'brown grease".[182] The amount of WCO generated from every country worldwide is huge and varies according to the amount of edible oil consumed, therefore, WCO is a readily available feedstock that can be used for biodiesel production.[181] The cost of WCO is generally lower than fresh edible oil as a major fraction contributing to the cost lies in the collection and purifying processes, for example the cost of WCO from soybean oil is expected to be 1.21 US$ per gallon by the year 2013 and the cost of soybean in the market is expected to rise to 2.47 US$ per gallon in 2013.[183] The physical and chemical properties of WCO are similar to their respective fresh edible oils. An increase in viscosity and specific heat, a change in the surface tension, color and the higher tendency of fat formation are reported.[184]

Heating value of WCO methyl ester is lower compared to petroleum-based diesel fuel, however CN and flash point are higher. It is also reported that the production of biodiesel from WCO is challenging due to the presence of undesirable components such as free fatty acids and water.[185] WCO can be a promising alternative as biodiesel feedstock.

8.4.5 Winery Waste (Grape Seeds)

World grape production was over 67 million metric tons in 2005, with Spain, France and Italy being the major world grape producers providing almost half of the total production.[186] Grapes are used in wine production and more than 20% of grape production typically becomes waste during production. Grape seed oil is obtained from the seeds left following pressing of the juice from grapes for wine making and seeds contain about 10–20% of oil, which is usually extracted with solvent and refined before use in order to remove unacceptable materials with the least possible loss of oil. The main characteristic of grape seed oil is its high content of unsaturated fatty acids, such as linoleic acid (72–76%, w/w).[187] Transesterification of the refined oil produced a biodiesel of good quality and showed good low-temperature flow properties.[186] Grape seed oil contains tocopherols, which can be used as antioxidants.

Table 8.1 Potential resources for biodiesel with their fatty acid compositions.

S. No.	Sources	Family	Common name	Myristic acid (C14:0)	Palmitic acid (C16:0)	Palmitoleic acid (C16:1)	Stearic acid (C18:0)	Oleic acid (C18:1)	Linoleic acid (C18:2)	Linolenic acid (C18:3)	Arachidic acid (C20:0)	Gadoleic acid (C20:1)
1.	Rhus succedanea Linn	Anacardiaceace	Wax tree	—	25.4	—	204.0	46.8	27.8	—	—	—
2.	Amona reticulata Linn	Annonaceae	Custard apple	1.0	17.2	4.2	7.5	48.4	21.7	—	—	—
3.	Ervatamia coronaria Stapf	Apocynaceae	East Indian rosebay, grape-jasmine	—	24.4	0.2	7.2	50.5	15.8	0.6	0.7	0.2
4.	Thevettia peruviana Merrill	Apocynaceae	Yellow oleander	—	15.6	—	10.5	60.9	5.2	7.4	0.3	—
5.	Vallaris solanacea Kuntze	Apocynaceae	Bread flower	—	7.2	—	14.4	35.3	40.4	—	1.8	—
6.	Balanites roxburghii Planch	Balanitaceae	Desert date	—	17.0	4.3	7.8	32.4	31.3	7.2	—	—
7.	Basella rubra Linn	Basellaceae	Ceylon spinach, Malabar spinach, Indian spinach	0.4	19.7	0.4	6.5	50.3	21.6	0.4	—	—
8.	Canarium commune Linn	Burseraceae	Elemi oil	—	29.0	—	9.7	38.3	21.8	1.2	—	—
9.	Celastrus paniculatus Linn	Celastraceae	Black-oil tree, intellect tree	—	25.1	—	6.7	46.1	15.4	3.0	—	—

Table 8.1 (*Continued*)

S. No.	Sources	Family	Common name	Myristic acid (C14:0)	Palmitic acid (C16:0)	Palmitoleic acid (C16:1)	Stearic acid (C18:0)	Oleic acid (C18:1)	Linoleic acid (C18:2)	Linolenic acid (C18:3)	Arachidic acid (C20:0)	Gadoleic acid (C20:1)
10.	*Euonymus hamiltonianuis* Wall	Celastraceae	Chinese spindle tree	—	18.3	—	1.5	39.1	25.8	5.3	—	—
11.	*Terminalia bellirica* Roxb	Combretaceae	Myrobalan, Hardad,	—	35.0	—	—	24.0	31	—	—	—
12.	*Terminalia chebula* Retz	Combretaceae	Myrobalan, Hardad, Chebulic Myrobalan	—	19.7	—	2.4	37.3	39.8	—	0.6	—
13.	*Vernonia cinerea* Less	Combretaceae	bitterleaf, Ewuro	3.0	23.0	—	8.0	32.0	22.0	—	3.0	—
14.	*Corylus avellana*	Corylaceae	Common hazel,	3.2	3.1	—	2.6	88.0	2.9	—	—	—
15.	*Croton tiglium* Linn	Euphorbaceae	Rushfoil, Croton	11.0	1.2	—	0.5	56.0	29.0	—	2.3	—
16.	*Jatropa curcas* Linn	Euphorbaceae	Barbados nut, Purging nut, Physic nut	—	14.2	1.4	6.9	43.1	34.4	—	—	—
17.	*Joannesia princeps* Vell	Euphorbaceae	Arara nut Tree, Cutieira,	2.4	5.4	—	—	45.8	46.4	—	—	—
18.	*Putranjiva roxburghii*	Euphorbaceae	Putranjiva, Lucky bean tree	—	8.0	—	15.0	56.0	18.0	—	3.0	—
19.	*Calophyllum apetalum* Wild	Guttiferae	Tamanu oil	—	8.0	—	14.0	48.0	30.0	—	—	—

Table 8.1 (*Continued*)

S. No.	Sources	Family	Common name	Myristic acid (C14:0)	Palmitic acid (C16:0)	Palmitoleic acid (C16:1)	Stearic acid (C18:0)	Oleic acid (C18:1)	Linoleic acid (C18:2)	Linolenic acid (C18:3)	Arachidic acid (C20:0)	Gadoleic acid (C20:1)
20.	*Calophyllum inophyllum* Linn	Guttiferae	Lexandrian laurel, Beauty leaf,	—	17.9	2.5	18.5	42.7	13.7	2.1	—	—
21.	*Garcinia combogia* Desr	Guttiferae	Garcinia kola, Malabar tamarind	—	2.3	—	38.3	57.9	0.8	0.4	0.3	—
22.	*Garcinia indica* Choisy	Guttiferae	Kokum	—	2.5	—	56.4	39.4	1.7	—	—	—
23.	*Mesua ferrea* Linn	Guttiferae	Ceylon ironwood, Indian rose chestnut	0.9	10.8	—	12.4	60.0	15.0	—	0.9	—
24.	*Nothapodytes nimmoniana* Milers	Icacinaceae	Ghanera	—	7.1	—	17.7	38.4	—	36.8	—	—
25.	*Illicium verum* Hook	Illiciceae	Star anise tree	4.43	—	—	7.93	63.24	24.4	—	—	—
26.	*Actinodaphne angustifolia*	Lauraceae	Pisa	1.9	0.5	—	—	5.4	—	—	—	—
27.	*Litsea glutinosa* Robins	Lauraceae	Indian laurel	—	—	—	—	2.3	—	—	—	—
28.	*Neolitsea cassia* Linn	Lauraceae	Grey bollywood, Smooth-barb bollygum	3.8	—	—	—	4.0	3.3	—	—	—

Table 8.1 (*Continued*)

S. No.	Sources	Family	Common name	Myristic acid (C14 : 0)	Palmitic acid (C16 : 0)	Palmitoleic acid (C 16 : 1)	Stearic acid (C18 : 0)	Oleic acid (C18 : 1)	Linoleic acid (C18 : 2)	Linolenic acid (C 18 : 3)	Arachidic acid (C 20 : 0)	Gadoleic acid (C20 : 1)
29.	*Neolitsea umbrosa* Gamble	Lauraceae	Chirandi	1.5	—	—	—	21.0	6.7	—	—	—
30.	*Michelia champaca* Linn	Magnoliaceace	Champa	—	20.7	6.9	2.5	22.3	42.5	—	2.6	—
31.	*Hiptage benghalensis* Kurz	Malpighiaceace	Helicopter flower, Hiptage	—	2.6	—	1.6	4.5	4.4	—	2.6	—
32.	*Aphanamixis polystachya* park	Meliaceae	Pithraj tree	—	23.1	—	12.8	21.5	29.0	13.6	—	—
33.	*Azadirachta indica*	Meliaceae	Neem, Lilac, Margosa tree	—	14.9	—	14.4	61.9	7.5	—	1.3	—
34.	*Melia azadirach* Linn	Meliaceae	Bead tree or Cape lilac	0.1	8.1	1.5	1.2	20.8	67.7	—	—	—
35.	*Swietenia mahagoni* Jacq	Meliaceae	Cuban mahogany tree, mahagni	—	9.5	—	18.4	56.0	—	16.1	—	—
36.	*Anamirta cocculus* Wight & Hrn	Menispermaceae	Fishberry or Levant nut	—	6.1	—	47.5	46.4	—	—	—	—
37.	*Broussonetid papyrifera* Vent	Moraceae	Paper mulberry	—	4.0	—	6.1	14.8	71.0	1.0	3.0	—
38.	*Moringa concanensis* Nimmo	Moringaceace	Konkan Moringa	—	9.7	—	2.4	83.8	0.8	—	3.3	—

Table 8.1 (*Continued*)

S. No.	Sources	Family	Common name	Myristic acid (C14 : 0)	Palmitic acid (C16 : 0)	Palmitoleic acid (C 16 : 1)	Stearic acid (C18 : 0)	Oleic acid (C18 : 1)	Linoleic acid (C18 : 2)	Linolenic acid (C 18 : 3)	Arachidic acid (C 20 : 0)	Gadoleic acid (C20 : 1)
39.	*Moringa oleifera* Lam	Moringaceace	Drumstick tree	—	9.1	2.1	2.7	79.4	0.7	0.2	5.8	—
40.	*Myristica malabarica* Lam	Myristicaceace	Malabar, Nutmeg	39.2	13.3	—	2.4	44.1	1.0	—	—	—
41.	*Argemone mexicana*	Papaveraceae	Mexican prickly poppy	0.8	14.5	—	3.8	18.5	61.4	—	1.0	—
42.	*Pongamia pinnata* Pierre	Papilionaceace	Pongam, Honge	—	9.8	—	6.2	72.2	11.8	—	—	—
43.	*Ziziphus mauritiana* Lam	Rhamnaceae	Jujube, Chinese apple, Indian plum	—	10.4	—	5.5	64.4	12.4	—	1.8	2.6
44.	*Princepia utilis* Royle	Rosaceace	Himalayan cherry	1.8	15.2	—	4.5	32.6	43.6	—	—	—
45.	*Meyna laxiflora* Robyns	Rubiaceae	Muyna	—	18.8	—	9.0	32.5	39.7	—	—	—
46.	*Nephelium lappaceum* Linn	Sapindaceace	Rambutan	—	2.0	—	13.8	45.3	—	—	34.7	4.2
47.	*Sapindus trifoliatus* Linn	Sapindaceace	South India soapnut	—	5.4	—	8.5	55.1	8.2	—	20.7	—
48.	*Madhuca butyracea* Mac	Sapotaceae	Indian butter tree	—	66.0	—	3.5	27.5	3.0	—	—	—
49.	*Madhuca indica* JF Gmel	Sapotaceae	Mahua	1.0	17.8	—	14.0	46.3	17.9	—	3.0	—

Table 8.1 (*Continued*)

S. No.	Sources	Family	Common name	Myristic acid (C14:0)	Palmitic acid (C16:0)	Palmitoleic acid (C16:1)	Stearic acid (C18:0)	Oleic acid (C18:1)	Linoleic acid (C18:2)	Linolenic acid (C18:3)	Arachidic acid (C20:0)	Gadoleic acid (C20:1)
50.	*Mimusops hexendra* Roxb	Sapotaceae	Bullet wood, Spanish cherry	—	19.0	—	14.0	63.0	3.0	—	1.0	—
51.	*Pterygota alata* Rbr	Sterculaceae	Buddha coconut	—	14.5	—	8.5	44.0	32.4	—	—	—
52.	*Holoptelia integrifolia*	Ulmaceae	Indian elm	3.5	35.1	1.9	4.5	53.3	—	—	1.1	—

8.5 Potential Resources

Azam *et al.* investigated the various resources of plants for biodiesel production.[5] On the basis of this, Table 8.1 lists 52 Indian plants (which contain more than 30% oil in their seed or fruit) with their major fatty acid compositions.[4,5,188–204] Some of these plants such as *Jatropha curcas*, *Pongamia pinnata* have already been discussed in detail earlier in this chapter.

References

1. G. Najafi, B. Ghobadian and T. F. Yusaf, *Renewable Sustainable Energy Rev.*, 2011, **15**, 3870.
2. A. L. Ahmad, N. H. Mat Yasin, C. J. C. Derek and J. K. Lim, *Renewable Sustainable Energy Rev.*, 2010, **15**, 584.
3. M. F. Demirbas, *Appl. Energy*, 2011, **88**, 3473.
4. A. Kumar and S. Sharma, *Renewable Sustainable Energy Rev.*, 2011, **15**, 1791.
5. M. M. Azam, A. Waris and N. M. Nahar, *Biomass Bioenergy*, 2005, **29**, 293.
6. G. A. Usher, *Dictionary of Plants used by Man*, Macmillan, London: Constable, 1974.
7. M. W. Pesman, *Meet Flora Mexicana*, Dale S. King, Arizona, 1962.
8. P. S. Ilavarasi, G. L. N. Rao, P. V. R. Iyer, K. Ravichandran and N. Rajendran, Narosa Publishing House, Delhi, p. 763.
9. N. V. Bringi, Non-traditional oilseeds and oils of India, Oxford & IBH Publishing, New Delhi, 1987.
10. B. R. Moser, *In Vitro Cell. Dev. Biol.–Plant*, 2009, **45**, 229.
11. J. B. Friday and D. Okano, *Calophyllum inophyllum* (kamani). in: Elevitch, C. R., ed. Traditional Trees of Pacific Islands. Permanent Agriculture Resources, Hōlualoa, HI, 2006, Pp. 183–198. http://www.traditionaltree.org
12. T. Said, M. Dutot, C. Martin, J. L. Beaudeux, C. Boucher, E. Enee, C. Baudouin, J-M. Warnet and P. Rat, *Eur. J. Pharm. Sci.*, 2007, **30**, 203.
13. P. K. Sahoo, L. M. Das, M. K. G. Babu and S. N. Naik, *Fuel*, 2007, **86**, 448.
14. A. C. Dweck and T. Meadowsy, *Int. J. Cosmet. Sci.*, 2002, **24**, 1.
15. http://www.svlele.com/undie.htm.
16. A. Johansson, *Biomass*, 1982, **2**, 103.
17. A. Karmakar, S. Karmakar and S. Mukherjee, *Bioresour. Technol.*, 2010, **101**, 7201.
18. S. Y. Lee, M. A. Hubbe and S. Saka, *Bioresources*, 2006, **1**, 150.
19. Y. Gaillard, A. Krishnamoorthy and F. Bevalot, *India. J. Ethnopharmacol.*, 2004, **95**, 123.
20. R. N. Chopra, S. L. Nayar and I. C. Chopra, Glossary of Indian medicinal plants Book, CSIR, New Delhi, 1956.
21. http://www.ntbg. org/plants/plant details.php?plantid=2601S.

22. R. N. Chopra, I. C. Chopra, R. L. Handa and D. L. Kapur, Indigenous drugs of India, Dhur & Sons, Calcutta, 1958.
23. B. Aliyu, B. Agnew and S. Douglas, *Biomass Bioenergy*, 2010, **34**, 1495.
24. G. Kafuku, H. Rutto and M. M. Mbarawa, presented at the 9th International Conference on Heat Engines and Environmental Protection, Budapest University of Technology and Economics, 2009.
25. R. Wang,, W. Zhou, M. A. Hanna, Y. Zhang, P. S. Bhadury, Y. Wang, B. Song and S. Yang, *Fuel*, 2012, **91**, 182.
26. L. R. G. Holm, J. Doll, E. Holm, J. Pancho and J. Herberger, John Wiley & Sons, New York, 1997, p. 273.
27. M. Friedman and C. E. Levin, *J. Agric. Food Chem.*, 1989, **37**, 998.
28. M. F. Ramadan, R. Zayed and H. El-Shamy, *Food Chem.*, 2007, **103**, 885.
29. H. Zhang, C. Han, M. Wang and Q. Yag, *Acta Bot. Boreali–Occident. Sin.*, 2008, **28**, 2538.
30. R. M. Blum, Jossey-Bass, Society and drugs, San Francisco, 1969, p. 122.
31. S. P. Agharkar, Medicinal plants Bombay presidency. Scientific Publishers, Jodhpur (India), 1991, p. 88.
32. S. K. Marwat, F. U. Rehman and S. Khan, *Gomal. Univ. J. Res.*, 2006, **22**, 4.
33. U. Rashid, F. Anwar and G. Knothe, *Fuel Process. Technol.*, 2009, **90**, 1157.
34. F. D. Gunstone and J. L. Harwood, *The Lipid Handbook*, CRC Press, Boca Raton, 3rd edn, 2007.
35. S. Ahmad, F. Anwar, A. I. Hussain and A. R. Awan, *J. Am. Oil Chem. Soc.*, 2007, **84**, 845.
36. A. S. Ramadhas, S. Jayaraj and C. Muraleedharan, *Fuel*, 2005, **84**, 335.
37. R. E. E. Jongschaap, W. J. Corré, P. S. Bindraban and W. A. Brandenburg, Claims and facts on Jatropha *curcas* L.: global Jatropha curcas evaluation. breeding and propagation programme, Plant Research International, Wageningen, 2007.
38. K. Openshaw, *Biomass Bioenergy*, 2000, **19**, 1.
39. B. N. Divakara, H. D. Upadhyaya, S. P. Wani, C. L. L. Gowda, *Appl. Energy*, **87**, 732.
40. R. P. S. Katwal and P. L. Soni, *Indian Forester*, 2003, **129**, 939.
41. http://pdf.usaid.gov/pdf docs/PNACH869.pdf
42. R. Singh, M. Kumar and E. Haider, *J. SAT Agri. Res.*, 2007, 5, 1.
43. D. Y. C. Leung, X. Wu and M. K. H. Leung, *Appl. Energy*, 2010, **87**, 1083.
44. S. Shah and M. N. Gupta, *Process Biochemistry*, 2007, **42**, 409.
45. C. Shanker and S. K. Dhyani, *Current Science*, 2006, **91**, 162.
46. S. V. Ghadge and H. Raheman, *Biomass Bioenergy*, 2005, **28**, 601.
47. V. Kumari, S. Shah and M. N. Gupta, *Fuel*, 2007, **21**, 368.
48. S. Tiwari, M. Saxena and S. K. Tiwari, *J. Appl. Polymer Sci.*, 2002, **87**, 110.

49. S. Puhan, N. Vedaraman, B. V. Ramabramhmam and G. Nagarajan, *J. Sci. Ind. Res.*, 2005, **64**, 890.
50. S. Puhan, N. Vedaraman and B. V. Ramabramhmam, in *Handbook of Plant Based Biofuels*, ed. A. Pandey, Taylor & Francis, Boca Raton, 2008.
51. Y. C. Bhatt, N. S. Murthy and R. K. Datta, *J. Inst. Eng. India Agri. Eng.*, 2004, **85**, 10.
52. C. E. Stavarache, J. Morris, Y. Maeda, I. Oyane, *Rev. Chim.*, 2008, **59**, 672.
53. A. Sarin, R. Arora, N. P. Singh, R. Sarin, M. Sharma and R. K. Malhotra, *J. Am. Oil Chem. Soc.*, 2010, **87**, 567.
54. P. T. Scott, L. Pregelj, N. Chen, J. S. Hadler, M. A. Djordjevic and P. M. Gresshoff, *Bioenergy Res.*, 2008, **1**, 2.
55. R. Kumar, M. Sharma, S. S. Ray, A. S. Sarpal, A. A. Gupta, D. K. Tuli, R. Sarin, R. P. Verma and N. R. Raje, SAE Publication No. 2004-28-0087, SAE International, Warrendale, 2004.
56. Government of India, *Troup's the Silviculture of Indian Trees*, Government of India Press, Nasik, 1983, vol. 4, p. 345.
57. L. C. Meher, S. N. Naik and L. M. Das, *J. Sci. Ind. Res.*, 2004, **63**, 913.
58. K. Sureshkumar, R. Velraj and R. Ganesan, *Renewable Energy*, 2008, **33**, 2294.
59. C. C. Akoh, S. W. Chang, G. C. Lee and J. F. Shaw, *J. Agric. Food Chem.*, 2007, **55**, 8995.
60. L. Wang and H. Yu, *Bioresour. Technol.*, 2012, **112**, 355.
61. J. Zhang, D. Fu, Z. Wei, H. Zhao and T. Zhang, *Acta. Ecol. Sin.*, 2006, **26**, 467.
62. E. P. Kris, P. S. Yu, J. Sabateé, H. Ratcliffe, G. Zhao and T. Etherton, *Nutr. Rev.*, 2001, **4**, 103.
63. S. K. Haldar, B. B. Ghosh and A. Nag, *Renewable Energy*, 2009, **34**, 343.
64. P. Rojas-Barros, A. D. Haro, J. Munoz and J. M. F. Martinez, *Crop Sci.*, 2004, **44**, 76.
65. J. A. Napier, *Ann. Rev. Plant Biol.*, 2007, **58**, 295.
66. http://www.svlele.com/jatropha plant.htm.
67. http://www.castoroil.in/crop/crop.html.
68. http://www.haryanaonline.com/Flora/ritha.htm
69. A. B. Chhetri, M. S. Tango, S. M. Budge, K. C. Watts and M. R. Islam, *Int. J. Mol. Sci.*, 2008, **9**, 169.
70. R. D. Misra and M. S. Murthy, *Fuel*, 2011, **90**, 2514.
71. A. B. Chhetri, K. C. Watts, M. S. Rahman and M. R. Islam, *Energy Resour. Part A Recovery Utilization Environ. Effects*, 2009, **31**, 1893.
72. Y. C. Sharma and B. Singh, *Fuel*, 2010, **89**, 1470.
73. T. Mamidi and N. A. Rawabawale, *Int. J Computer Information System*, 2012, **4**, 72.
74. N. Vedaramana, S. Puhan, G. Nagarajan, B. V. Ramabrahmam and K. C. Velappan, *Ind. Crops Prod.*, 2012, **36**, 282.

75. http://www.ijec.net.
76. M. S. Radwan, S. K. Selim and A. M. Kader, in *Proc. SAE*, SAE International, Warrendale, 1997.
77. M. Al-Muhtaseb, BSc Thesis, Jordan university of science and technology, 2005.
78. H. Al-Zoubi, MSc Thesis, Jordan university of science and technology, 1996.
79. M. K. Abu-Arabi, M. A. Allawzi, H. S. Al-Zoubi and A. Tamimi, *Chem. Eng. J.*, 2000, **76**, 61.
80. J. Busson-Breysse, M. Farines and J. Soulier, *J. Am. Oil Chem. Soc.*, 1994, **71**, 999.
81. A. Shani, *Chemtech.*, 1995, **25**, 49.
82. H. Lorenzi, Árvores Brasileiras: manual de identificac̨ão e cultivo de plantas arbóreas nativas do Brasil. Editora Plantarum Ltda, Nova Odesa. São Paulo, Brasil, 1992.
83. Z. N. Diniz, P. S. Bora, V. Q. Neto and J. M. O. Cavalheiro, *Grasas Aceites*, 2008, **59**, 160.
84. M. H. Chaves, A. S. Barbosa, N. J. M. Moita, S. A. Pimentel and J. H. G. Lago, *Quím. Nova*, 2004, **27**, 404.
85. R. Sarin, M. Sharma and A. A. Khan, *Bioresour. Technol.*, 2010, **101**, 1380.
86. K. K. Anand, B. Singh, A. K. Saxena, B. K. Chandan, V. N. Gupta and V. Bhardwaj, *Pharmacol. Res.*, 1997, **36**, 315.
87. K. M. Elizabeth, *Indian J. Clin. Biochem.*, 2005, **20**, 150.
88. U. Rashid, F. Anwar and G. Knothe, *Biomass Bioenergy*, 2011, **35**, 4034.
89. W. L. Wagner, D. R. Herbst and S. H. Sohmer, *Bernice Pauahi Bishop Museum Special Publication*, University of Hawaii Press, Honolulu, vol. 83, 1990.
90. R. S. Subbaram, *Proc. Indian Acad. Sci.*, 1954, **39**, 301.
91. R. Ilavarasan, M. Vasudevan, S. Anbazhagan and S. Venkataraman, *J. Ethnopharmacol.*, 2003, **87**, 227.
92. T. Balusamy and R. Manrappan, *J. Sci. Ind. Res.*, 2007, **66**, 1035.
93. S. A. Ibiyemi, S. S. Bako, G. O. Ojukuku and V. O. Fadipe, *J. Am. Oil Chem. Soc.*, 1995, **72**,745.
94. S. A. Ibiyemi, V. O. Fadipe, O. O. Akinremi and S. S. Bako, *J. Appl. Sci. Environ. Manage.*, 2002, **6**, 61.
95. Y. Chen, J. Chen and Y. Luo, *Renewable Energy*, 2012, **44**, 305.
96. H. A. M. van der Vossen and G. S. Mkamilo (Editors), *Plant Resources of Tropical Africa*, 14. Vegetable oils. PROTA Foundation, Wageningen, Netherlands/Backhuys Publishers, 2007.
97. S. Gryglewicz, K. Grabas and G. Gryglewicz, *Bioresour. Technol.*, 2000, **75**, 213.
98. J. Y. Park, D. K. Kim, Z. M. Wang, P. Lu, S. C. Park and J. S. Lee, *Appl. Biochem. Biotechnol.*, 2008, **148**, 109.
99. Y. H. Chen, J. H. Chen, C. Y. Chang and C. C. Chang, *Bioresour. Technol.*, 2010, **101**, 9521.

100. V. Sharma and P. P. Kundu, *Prog. Polym. Sci.*, 2006, **31**, 983.
101. P. D. Tyagi and K. K. Kakkar, Non-conventional vegetable oils, Batra Book Service, New Delhi, 1991.
102. J. P. Davis, L. O. Dean, W. H. Faircloth and T. H. Sanders, *J. Am. Oil Chem. Soc.*, 2008, **85**, 235.
103. http://www.nass.usda.gov.
104. S. F. O'Keefe, V. A. Wiley and D. A. Knauft, *J. Am. Oil Chem. Soc.*, 1993, **70**, 489.
105. C. Kaya, C. Hamamci, A. Baysal, O. Akba, S. Erdogan and A. Saydut, *Renewable Energy*, 2009, **34**, 1257.
106. R. A. Holser and R. H. O'Kuru, *Fuel*, 2006, **85**, 2106.
107. P. C. Crews, S. A. Sievert, L. T. Woeppel and E. A. McCullough, *Text Res. J.*, 1991, **61**, 203.
108. E. E. Gaertner, *Econ. Bot.*, 1979, **33**, 119.
109. R. P. Adams, M. F. Balandrin and J. R. Martineau, *Biomass*, 1984, **4**, 81.
110. B. P. Chapagain, Y. Yehoshua and Z. Wiesman, *Bioresour. Technol.*, 2009, **100**, 1221.
111. J. B. Hall and D. H. Walker, *B. Aegyptiaca Del. – A Monograph*. School of Agricultural and Forest Science, University of Wales, Bangor, 1991.
112. M. Cardone, M. Mazzoncini, S. Menini, V. Rocco, A. Senatore, M. Seggiani and S. Vitolo, *Biomass Bioenergy*, 2003, **25**, 623.
113. http://www.biodieselmagazine.com/articles/2063/count-on-canola-for-your-biodiesel.
114. J. Park, D. Kim J. Lee, S. Park, Y. Kim and J. Lee, *Bioresour. Technol.*, 2008, **99**, 1196.
115. http://www.openmarket.org.
116. B. R. Moser, *Fuel*, 2012, **92**, 231.
117. Y. X. Xu and M. A. Hanna, *Ind. Crops Prod.*, 2007, **26**, 69.
118. Y. X. Xu and M. A. Hanna, *Ind. Crops Prod.*, 2009, **29**, 473.
119. U. Rashid, H. A. Rehman, I. Hussain, M. Ibrahim and M. S. Haider, *Energy*, 2011, **36**, 5632.
120. A. M. Khushk and M. L. Lashari, *J. Agri. Res.*, 2007, **32**, 359.
121. J. Hemavatahy, *J. Food Compos. Anal.*, 1992, **5**, 90.
122. Y. Hu Mian-hao Ao, *Int. J. Food Sci. Technol.*, 2007, **42**, 1397.
123. N. A. M. Yanty, O. M. Lai, A. Osman, K. Long and H. M. Ghazali, *J. Food Lipids*, 2008, **15**, 42.
124. Y. Basiron, *Eur. J. Lipid Sci. Technol.*, 2007, **109**, 289.
125. M. K. Lam, K. T. Tan, K. T. Lee and A. R. Mohamed, *Renewable Sustainable Energy Rev.*, 2009, **13**,1456.
126. R. P. Singh, M. H. Ibrahim, N. Esa and M. S. Iliyana, *Rev. Environ. Sci. Biotechnol.*, 2010, **9**, 331.
127. D. O. Edem, *Plant Foods Hum. Nutr.*, 2002, **57**, 319.
128. K. Y. Foo and B. H. Hameed, *Renewable Sustainable Energy Rev.*, 2009, **13**, 2495.
129. J. Guo and A. C. Lua, *Biomass Bioenergy*, 2001, **20**, 223.

130. http://journeytoforever.org.
131. http://www.ufop.de.
132. D. Pimental and W. T. Patzek, *Natural Resources Res.*, 2005, **14**, 65.
133. R. S. Martín, T. de la Cerda, A. Uribe, P. Basilio, M. Jordán, D. Prehn and M. Gebauer, *Fuel*, 2010, **89**, 3785.
134. F. A. Barkley, *Lilloa.*, 1957, **28**, 111.
135. D. S. Seigler, M. Cortes and J. M. Aguilera, *Biochem Syst. Ecol.*, 1987, **15**, 71.
136. J. Van Gerpen, B. Shanks, R. Pruszko, D. Clements and G. Knothe, Report No. SR-510-36244National Renewable Energy Laboratory, Golden, Colorado, USA, 2004.
137. K. W. C. Sze-Tao and S. K. Sathe, *J. Sci. Food Agric.*, 2000, **80**, 1393.
138. Foreign Agricultural Service, United States Department of Agriculture, Tree nuts: world markets and trade. Trade report Washington, D.C. USA, 2010. http://www.fas.usda.gov/htp/horticulture/Tree%20Nuts/2010_10_TreeNuts.pdf
139. G. P. Savage, P. C. Dutta and D. L. McNeil, *J. Am. Oil Chem. Soc.*, 1999, **76**, 1059.
140. U. Rashid, F. Anwar, B. R. Moser and G. Knothe, *Bioresour. Technol.*, 2008, **99**, 8175.
141. J. F. Morton, *Econ. Bot.*, 1991, **45**, 318.
142. M. C. Palada and L. C. Changl, in *International Cooperators' Guide*, No. 03-545, pp. 1–5, Asian Vegetable Research and Development Centre, Shanhua, Taiwan, 2003.
143. Council of Scientific and Industrial Research, Council of Scientific and Industrial Research, The Wealth of India: Raw Materials, vol. 6, 1962, p. 425, New Delhi.
144. A. Sengupta and M. P. Gupta, *Fette, Seifen, Anstrichm.*, 1970, **72**, 6.
145. S. Lalas and J. Tsaknis, *J. Am. Oil Chem. Soc.*, 2002, **79**, 677.
146. F. Anwar, S. Latif, M. Ashraf and A. H. Gilani, *Phytother. Res.*, 2007, **21**, 17.
147. S. Einloft, T. O. Magalhaes, A. Donato, J. Dullius and R. Ligabue, *Energy Fuels*, 2008, **22**, 671.
148. S. Sinha, A. K. Agarwal and S. Garg, *Energy Convers. Manage.*, 2008, **49**, 1248.
149. Y. H. Ju and S. R. Vali, *J. Sci. Ind. Res.*, 2005, **64**, 866.
150. N. Kumar, *J. Sci. Ind. Res.*, 2007, **66**, 399.
151. J. Huang, Y.Wang, J. Qin and A. P. Roskilly, *Fuel Process. Technol.*, 2010, **91**, 1761.
152. http://efloras.org/florataxon.aspx?flora_id=2&taxon_id=200012703.
153. http://ec.europa.eu/research/energy/pdf/41_tao_wang_en.pdf.
154. A. Saydut,, M. Z. Duz, C. Kaya, A. B. Kafadar and C. Hamamci, *Bioresour. Technol.*, 2008, **99**, 6656.
155. M. Elleuch, S. Besbes, O. Roiseux, C. Blecker and H. Attia, *Food Chem.*, 2007, **103**, 641.

156. H. Koca, M. Bor, F. Ozdemir and I. Turkan, *Environ. Exp. Bot.*, 2007, **60**, 344.

157. K. T. da Silva de La Salles, S. M. P. Meneghettib, W. F. de La Salles, M. R. Meneghetti, I. C. F. dos Santos, J. P. V. da Silva, S. H. V. de Carvalho and J. I. Soletti, *Ind. Crops Prod.*, 2010, **32**, 518.

158. J. Li, Y. Fu, X. Qu, W. Wang, M. Luo, C. Zhao and Y. Zu, *Bioresour. Technol.*, 2012, **108**, 112.

159. Y. Li, B. Wang, N. Wu and C. Q. Lan, *Appl. Microbiol. Biotechnol.*, 2008, **81**, 629.

160. Y. Li, M. Horsman, N. Wu, C. Q. Lan and N. Dubois-Calero, *Biotechnol. Prog.*, 2008, **24**, 815.

161. M. N. Campbell, *Guelph Eng. J.*, 2008, **1**, 2.

162. A. Richmond, *Handbook of Microalgal Culture: Biotechnology and Applied Phycology*, Blackwell Science, Oxford, 2004.

163. A. Singh, N. P. Singh and J. D. Murphy, *Bioresour. Technol.*, 2011, **102**, 26.

164. J. Singh and S. Gu, *Renewable Sustainable Energy Rev.*, 2010, **14**, 2596.

165. J. Janaun and N. Ellis, *Renewable Sustainable Energy Rev.*, 2010, **14**, 1312.

166. P. T. Vasudevan and M. Briggs, *J. Ind. Microbiol. Biotechnol.*, 2008, **35**, 421.

167. M. K. Danquah, B. Gladman, N. Moheimani and G. M. Forde, *Chem. Eng. J.*, 2009, **151**, 73.

168. K. M. McGinnis, T. A. Dempster and M. R. Sommerfeld, *J. Appl. Phycol.*, 1997, **9**, 19.

169. L. Rodolfi, G. C. Zittelli, N. Bassi, G. Padovani, N. Biondi, G. Bonini and M. R. Tredici1, *Biotechnol. Bioeng.*, 2009, **102**, 100.

170. A. Vonshak, *Biotechnol. Adv.*, 1990, **8**, 709.

171. J. Lu, K. Nie, F. Xie, F. Wang and T. Tan, *Process Biochem.*, 2007, **42**, 1367.

172. C. Oner and S. Altun, *Appl. Energy*, 2009, **86**, 2114.

173. Z. Wen and M. B. Johnson, Publication No. 442-886, Virginia Polytechnic Institute and State University, Blacksburg, 2009.

174. M. Guru, B. D. Artukoglu, A. Keskin and A. Koca, *Energy Convers. Manage.*, 2009, **50**, 498.

175. M. Sheedlo, *Basic Biotechnol.*, 2008, **4**, 61.

176. E. Uggetti, E. Llorens, A. Pedescoll, I. Ferrer, R. Castellnou and J. Garcia, *Bioresour. Technol.*, 2009, **100**, 3882.

177. E. Jarde, L. Mansuy and P. Faure, *Water Res.*, 2005, **39**, 1215.

178. S. Dufreche, R. Hernandez, T. French, D. Sparks, M. Zappi, E. Alley, *J. Am. Oil Chem. Soc.*, 2007, **84**,181.

179. http://www.london.ca/d.aspx?s=/Sewer and Wastewater/ Sewagetreatment index.htm

180. F. Smesdes and T. K. Askaland, *Mar. Pollut. Bull.*, 1999, **38**, 193.

181. M.M. Gui, K.T. Lee and S. Bhatia, *Energy*, 2008, **33**, 1646.

182. M. G. Kulkarni and A. K. Dalai, *Ind. Eng. Chem. Res.*, 2006, 45, 2901.

183. http://www.eia.doe.gov/oiaf/analysispaper/biodiesel/index.html.

184. K. Jacobson, R. Gopinath, L. C. Meher and A. K. Dalai, *Appl. Catal. B. Environ.*, 2008, **85**, 86.
185. C. M. Fernández, M. J. Ramos, Á. Pérez and J. F. Rodríguez, *Bioresour. Technol.*, 2010, **101**, 7019.
186. M. Martinello, G. Hecker, M. C. Pramparo, *J. Food Eng.*, 2007, **81**, 60.
187. Publication & Information Directorate, Council of Scientific & Industrial Research, New Delhi, The Wealth of India: A Dictionary of Indian Raw Materials & Industrial Products, vol. 9, 1972.
188. Publication & Information Directorate, Council of Scientific & Industrial Research, New Delhi, The Wealth of India: A Dictionary of Indian Raw Materials & Industrial Products, (revised), vol. 1, 1985.
189. A. Sarin, R. Arora, N. P. Singh, R. Sarin, R. K. Malhotra and K. Kundu, *Energy*, 2009, **34**, 2016.
190. V. K. Saxena and S. K. Jain, *Fitoterapia*, 1990, **61**, 348.
191. A. Ghanim, *Indian Farming*, 1991, **41**, 9.
192. Publication & Information Directorate, Council of Scientific & Industrial Research, New Delhi, The Wealth of India: A Dictionary of Indian Raw Materials & Industrial Products, (revised), vol. 2, 1988.
193. T. P. Hilditch and P. N. Williams, *The Chemical Constituents of Natural Fats*, Chapman and Hall, London, 4th edn, 1964.
194. A. Sengupta, S. C. Sengupta and U. K. Mazumdar, *Fett. Wissenschaft Technol.*, 1987, **89**, 119.
195. Publication & Information Directorate, Council of Scientific & Industrial Research, New Delhi, The Wealth of India: A Dictionary of Indian Raw Materials & Industrial Products, vol. 6, 1962.
196. Publication & Information Directorate, Council of Scientific & Industrial Research, New Delhi, The Wealth of India: A Dictionary of Indian Raw Materials & Industrial Products (revised), vol. 3, 1992.
197. Publication & Information Directorate, Council of Scientific & Industrial Research, New Delhi, The Wealth of India: A Dictionary of Indian Raw Materials & Industrial Products, vol. 5, 1959.
198. Publication & Information Directorate, Council of Scientific & Industrial Research, New Delhi, The Wealth of India: A Dictionary of Indian Raw Materials & Industrial Products, vol. 7, 1966.
199. R. Banerji, S. C. Verma and P. Pushpendra, *Natural Product Radiance*, 2003, **2**, 68.
200. A. Bhattacharjee, S. K. Ghosh, D. Ghosh, S. Ghosh, M. K. Maiti and S. K. Sen, *Plant Sci.*, 2002, **163**, 791.
201. A. Singh and I. S. Singh, *Food Chem.*, 1991, **40**, 221.
202. Publication & Information Directorate, Council of Scientific & Industrial Research, New Delhi, The Wealth of India: A Dictionary of Indian Raw Materials & Industrial Products, vol. 10, 1976.
203. http://www.flowersofindia.net/botanical.html.

CHAPTER 9

Present State and Policies of the Biodiesel Industry

9.1 Introduction

Biodiesel production depends upon governmental policies. This chapter highlights the present state and policies related to the biodiesel industries of different countries. The EU, the USA, India and various other countries are covered.

9.2 The European Union

The European Biodiesel Board (EBB) is a non-profit organization established in January 1997 to see the promotion of biodiesel in the EU. In the EU biodiesel is by far the biggest biofuel and represents 82% of biofuel production. Biodiesel production uses around 1.4 million hectares (ha) of arable land in the EU.[1]

9.2.1 EU Biodiesel Production and Capacity Leaders

Biodiesel production is increasing year-on-year in the EU (Figure 9.1).[2] Biodiesel production increased from 1065 million metric tons (MMT) in 2002 to 9570 MMT in 2010.

Recently published EBB biodiesel production and capacities statistics showed that in 2010, EU biodiesel production registered a 5.5% increase compared to the previous year, reaching a level of 9.57 MMT, but the forecast for 2011 shows a reduction in EU production compared to the same time last

Biodiesel: Production and Properties
Amit Sarin
Published by the Royal Society of Chemistry, www.rsc.org

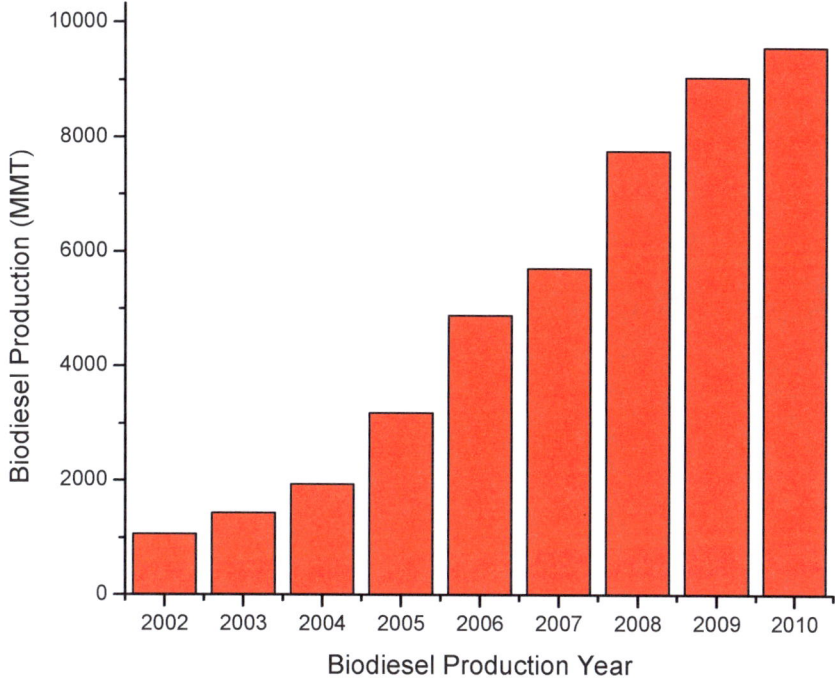

Figure 9.1 Biodiesel production in the EU.

year (Table 9.1).[2] In addition, the 2010 increase remains low compared to growth rates registered in 2009 (17%) and 2008 (35%). The first semester of 2011 shows a reduction in domestic EU production. The number of existing biodiesel facilities stood at 254, slightly up from 2009 due to the start of a few new production units. This industrial basis is the result of investments in biodiesel production planned before 2007 in order to meet the ambitious objectives for biofuel consumption set by the EU authorities.[3]

In 2010, Germany and France remained by far the leading biodiesel-producing nations, and Spain confirmed its position of third leading EU biodiesel producer, ahead of Italy, which saw a slight decline in production compared to 2009. The UK is in fifteenth position with a marginal increase in biodiesel production in comparison to 2009, but there was decrease in production capacity from 609 million tons in 2010 to 404 million tons (MT) in 2011 (Table 9.2).[2]

Pure biodiesel use is predominant in Germany. In Germany, the current program of development of the biodiesel industry is not a special exemption from EU law, but rather is based on a loophole in the law. The motor fuels tax in Germany is based on mineral fuel. Since biofuel is not a mineral fuel, it can be used for motor transport without being taxed. In France, biodiesel production started in 1992. In contrast to Germany, French biodiesel is exclusively sold as a blend with either up to 5% or up to 30% biodiesel added to

Table 9.1 EU 2009 and 2010 biodiesel production estimates.[a]

Country	2009 Biodiesel production/ MMT	2010 Biodiesel production/ MMT
Germany	2539	2861
France	1959	1910
Spain	859	925
Italy	737	706
Belgium	416	435
Poland	332	370
Netherlands	323	368
Austria	310	289
Portugal	250	289
Denmark	233	246
Sweden	233	246
Finland[b]	220	288
Czech Republic	164	181
Hungry	133	149
UK	137	145
Slovakia	101	88
Lithuania	98	85
Romania	29	70
Latvia	44	43
Greece	77	33
Bulgaria	25	30
Ireland[b]	17	28
Slovenia	9	22
Cyprus	9	6
Estonia	24	3
Malta	1	0
Luxemburg	0	0
Total	9046	9570

[a]Subject to ± 5% error. [b]Data include hydrodiesel

fossil diesel.[4] Biodiesel was introduced in 1993 in Germany and it is the leading producer and consumer of biodiesel in the EU. The National Government in Germany has been promoting biofuels in particular and bioenergy in general since the early 1990s because bioenergy is not only viewed as an important energy source for environmental reasons and energy security, but also as a way to support the agricultural industry and regional development.[5] Most of the biodiesel is produced domestically with some imports and only negligible exports.[6] Until 2004, biodiesel was only utilized as B100 by some car users, truck operators and bus fleets. Since 2004, the excise duty exemption was extended to low-level blends, such as B5, thereby stimulating oil companies to initiate blending. The National Government in collaboration with oil companies and automobile manufacturers, as well as research institutes in Germany, have formulated a transport fuels strategy, which explicitly addresses biofuels.[5] Starting from 1 January 2007, Germany eliminated fuel

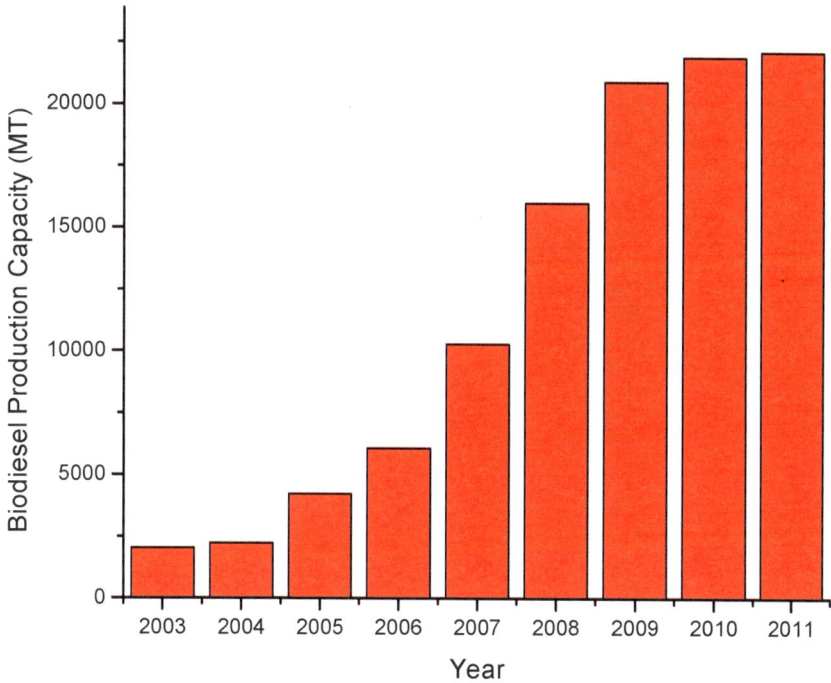

Figure 9.2 Biodiesel production capacity in the EU.

excise tax exemptions and replaced them with quota obligations and tax rebates.

Biodiesel production capacity is also increasing year-on-year in the EU (Figure 9.2).[2] Biodiesel production capacity increased from 2048 MT in 2003 to 22 117 MT in 2011. Regarding EU biodiesel production capacity, as of July 2011, it had reached more than 22 MT (Table 9.2).[2]

While EU biodiesel capacity utilization currently stands at 44%, for the first two quarters of 2011, and for the first time in registered history, the entire EU production has slightly decreased year-on-year.[2] The reason is increased imports from other countries such as Argentina and Indonesia as well as circumvention measures from North America.

9.2.2 The Renewable Energy Directive

The general EU policy rotates about three objectives: (a) competitiveness of the EU economy, (b) security of energy supply, and (c) environmental protection. The Renewable Energy Directive 2009/28 (RED) adopted in December 2008 is further step. The RED was designed to create a strong framework for the development of the biofuels industry in the EU and took a decision to introduce a 10% binding target for renewable energy use in transport. However, Member States' progress in implementing the RED has

Table 9.2 EU 2010 and 2011 biodiesel production capacity estimates.

Country	2010 Production capacity/MT	2011 Production capacity/MT
Germany	4933	4932
France	2505	2505
Spain	4100	4410
Italy	2375	2265
Belgium	670	710
Poland	710	864
Netherlands	1328	1452
Austria	560	560
Portugal	468	468
Denmark	250	250
Sweden	277	277
Finland[a]	340	340
Czech Republic	427	427
Hungry	158	158
UK	609	404
Slovakia	156	156
Lithuania	147	147
Romania	307	277
Latvia	156	156
Greece	662	802
Bulgaria	425	348
Ireland[a]	76	76
Slovenia	105	113
Cyprus	20	20
Estonia	135	135
Malta	5	5
Luxemburg	0	0
Total	21 904	22 117

[a]Data include hydrodiesel.

been limited and scattered, showing the need for greater harmonization at EU level. Efforts must be accelerated to finalize the RED implementation in 2012. The implementation of the 'double-counting' mechanism for biofuels produced from waste and residues requires particular attention. Reports have suggested that if not consistently implemented in Member States' legislations, the double-counting mechanism will inevitably lead to important disruptions of the EU biofuels market, while completely undermining the objectives of the RED regarding the sustainability of biofuels.[2]

The EBB believes that two set of actions should be urgently developed by EU and national authorities, the first is defining a robust monitoring and traceability mechanism that will allow verification of the validity of claims for double-counting biofuels based on an EU-wide traceability scheme; and the second is clarifying what types of biofuels will be double-counted, in particular by clearly defining what materials should be considered as 'residues'.[2]

9.2.3 Debate of Indirect Land Use Change.

The EBB has rejected the findings of an EU-funded study into the indirect land use change (ILUC) effect of biofuel production as unscientific. The EBB rejected a report published in March 2010 by the International Food Policy Research Institute (IFPRI). It rejected the findings of a draft EU study showing that the cultivation of rapeseed to make road transport fuels is worse for the climate than using conventional diesel. The EBB indicates that the IFPRI study cannot serve as a basis for policy-making on the highly debatable and unscientific concept of ILUC. By relying on a unique piece of research, conducted by the US-based consultancy IFPRI, a knee-jerk reaction by EU politicians could wipe out the EU's 17.5 US$ billion biofuel production industry overnight and play into the hands of biofuel makers in Asia and South America.[7] In the view of EBB, it is essential that efforts to implement the RED and its 2020 objectives are not diverted by the current debate over biofuels ILUC. "Can an industry like the biodiesel industry be the number one renewable fuel industry in the EU or be at risk of closing its production plants because of something that is not validated?" said EBB Secretary General Raffaello Garofalo at a news conference in Brussels.

A report has mentioned that, the current debate on ILUC completely overlooks the fact that the RED currently represents the most comprehensive and stringent set of sustainability criteria applying to biofuel production, already guaranteeing that only biofuels with a high sustainability profile are placed on the EU market. Ultimately, land-use change is an issue that has to be addressed at a global level, *i.e.*, looking at the range of sectors having an impact on land use worldwide and not only at biofuels.[3]

9.2.4 Future Plan

In April 2009 the parliament of the EU endorsed a minimum binding target of 10% for biofuels in transport by 2020 as part of the RED. The RED also specified a minimum 35% reduction in greenhouse gas (GHG) emissions to be achieved by biofuels during their lifecycle, a target that is meant to increase to at least 50% from 2017. Sustainability criteria for ILUC changes are also provided. No bio-feedstock shall originate from primary forests, highly bio-diverse grasslands, protected territories or carbon-rich areas.[2]

As mentioned earlier, Argentina and Indonesia represented respectively 71 and 27% of EU biodiesel imports. Both of these countries maintain differential export tax (DET) regimes that artificially incentivize the exports of biodiesel instead of soybean and palm oil. It should be reminded that the Argentinean DET scheme is strongly distorting trade as it maintains a large differential between an export tax on crude soybean oil of 32% and an export tax on biodiesel of only 20% (the effectively applied rate is in fact 14,16% Tax rate), therefore incentivizing the export of the finished product biodiesel. This concept applies also to Indonesian DETs.[2]

For this reason, the EBB welcomed the recent Commission proposal to remove Argentina and Malaysia from the list of countries benefiting from the EU Generalized System of Preferences (GSP) and if adopted by the Council and the Parliament, this proposal will at least ensure that Argentine and Malaysian biodiesel will be subject to the non-preferential import duty of 6.5%, instead of benefiting from duty-free access to the EU market.[2]

9.3   The USA

In his 2012 state of the Union address, President Barack Obama stated that: "This country needs an all-out, all-of-the-above strategy that develops every available source of American energy—a strategy that's cleaner, cheaper, and full of new jobs".[8] He further stated that "I am directing my administration to allow the development of clean energy on enough public land to power three million homes. And I'm proud to announce that the Department of Defense, the world's largest consumer of energy, will make one of the largest commitments to clean energy in history—with the Navy purchasing enough capacity to power a quarter of a million homes a year".

The National Biodiesel Board (NBB) has issued a response to the speech. "The US biodiesel industry is proving that we can accomplish the president's goals of creating jobs while building a clean-energy economy", said Anne Steckel, NBB's vice president of federal affairs. "With the help of strong domestic energy policy, we had a record year of production last year and supported nearly 40 000 jobs across the country".[8]

Enzyme company Novozymes has also spoken out in support of the speech: "It's proven that home-grown, renewable energy can put steel in the ground, create jobs and power our economy", said Adam Monroe, President of Novozymes, North America. "Working with the President, we can help America become less dependent on foreign oil and a smarter consumer of energy. Innovations like advanced biofuels can play a major role in the President's vision but we need steady policies like the Renewable Fuel Standard—and we look forward to working with Congress to preserve them".[8]

9.3.1   The Energy Policy Act Announced Record Biodiesel Production in 2011

In the USA, the Energy Policy Act (EPA) of 2005 was one of the most significant steps.[9] The EPA is a 14 US$ billion national energy plan containing numerous provisions related to energy efficiency and conservation, modernization of energy infrastructure, and promotion of both traditional energy sources and renewable alternatives. It contains many provisions to spur development of biofuels. Title IX (pertaining to R&D) includes several sections encouraging collaboration between government, industry, and academic institutions to develop advanced technologies for the production of biofuels. This title also includes production incentives for cellulosic biofuels derived from non-edible plant material, to ensure that an annual production of 1 billion gallons per year is achieved by 2015. Biodiesel is the first and only commercial-scale fuel used across the USA to meet the EPA's definition as an advanced biofuel. It is a renewable, clean-burning diesel replacement that can be used in existing diesel engines and meets strict ASTM fuel specifications. Made from an increasingly diverse mix of resources such as agricultural oils, recycled cooking oil and animal fats, it is produced in nearly every state in the country.

According to year-end numbers released by the EPA on 27 January 2012, the USA biodiesel industry reached a key milestone by producing more than 1 billion gallons of fuel in 2011 and it is by far a record for the industry and easily exceeds the 800 million gallon target required under the EPA's renewable fuel standard (RFS) (Figure 9.3).[10,11] A recent economic study commissioned by the NBB found that biodiesel production of 1 billion gallons supports 39 027 jobs across the country and more than 2.1 US$ billion in household income and an additional 11 698 jobs could be added between 2012 and 2013 alone under continued growth in the RFS and with an extension of the biodiesel tax incentive. The biodiesel industry's success in 2011 comes after Congress reinstated the fuel's 1 US$ per gallon tax credit in December 2010 and as the EPA's RFS program for biodiesel completed its first full year of implementation because without those policies in place in 2010, production

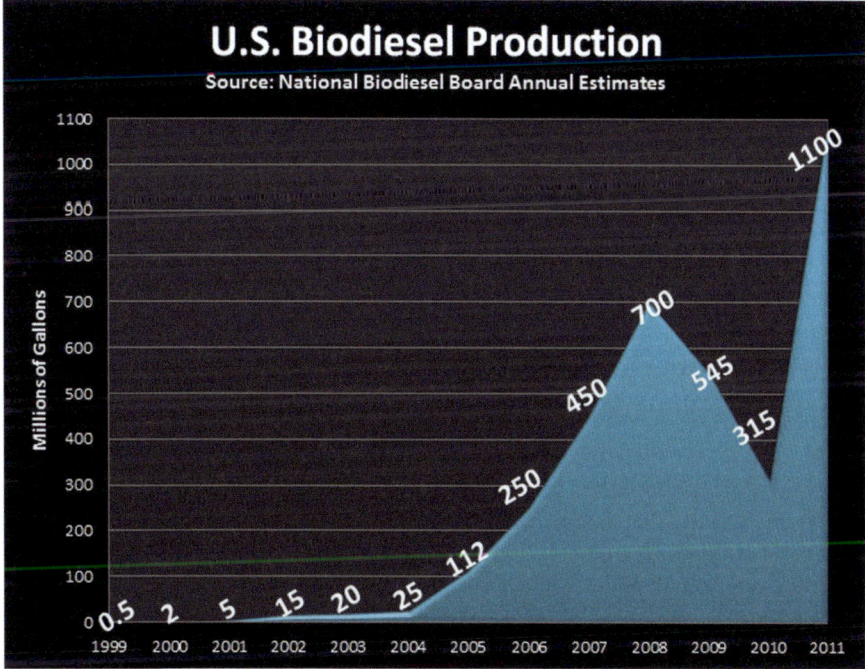

Figure 9.3   USA biodiesel production in 2011.

dropped dramatically as dozens of plants shut down and thousands of people lost jobs.[10] The latest EPA numbers show that a record 160 million gallons of biomass-based diesel were produced in December alone.

9.3.2 Why Biodiesel Production Slipped in Early 2012

According to new numbers released by the EPA, the USA biodiesel industry produced 135 million gallons of fuel in the months January and February of 2012. The volume is an increase over the same period last year, when production totaled less than 80 million gallons, but it is down from the record production late last year when the industry exceeded 100 million gallons per month for five consecutive months and reached a peak of 160 million gallons in December.[12]

"These are solid numbers that show the biodiesel industry is on pace to meet the 1 billion gallon RFS requirement this year, but they also reflect some of the missed opportunities for growth and jobs that we've seen with the loss of the tax credit and the continued uncertainty about next year's RFS volume" Anne Steckel, vice-president of federal affairs for the NBB said. "With the tax credit and clear RFS growth in place, we think these numbers would be better".

The tax incentive expired on 31st December 2011 and the biodiesel industry is urging Congress to reinstate it. In addition, the industry is calling on the EPA and the Obama Administration to finalize the EPA's proposal to boost the biodiesel volume requirement under the RFS to 1.28 billion gallons in 2013.[10]

9.3.3 Biodiesel Policy Trends in the USA

The USA Department of Transportation (DOT) is funding 46 innovative transit projects aimed at helping reduce the nation's dependence on oil while creating a marketplace for green jobs.[13] On 17th November 2011, Transportation Secretary Ray LaHood announced that these projects will share a combined 112 US$ million in funding from the Federal Transit Administration. Information released by the DOT shows three biodiesel projects will receive funding through this round of grants under the Clean Fuels Program: first, the city of Gainesville, has been awarded 3 US$ million to produce high-efficiency biodiesel buses and electric cooling system conversions for existing biodiesel buses; second, the Intercity Transit in Thurston County has been awarded 1.5 US$ million to purchase hybrid biodiesel–electric replacement buses; third, the city of Longview has been awarded 1.1205 US$ million to purchase 35-foot clean fuel biodiesel buses.

Last year, Agriculture Secretary Tom Vilsack announced that the United States Department of Agriculture (USDA) will make payments to more than 160 biofuel producers in 41 states under the Bioenergy Program for Advanced Biofuels.[14] Ever Cat Fuels is one of the many biodiesel producers to receive payments under the 9005 program. According to the USDA, the company has been selected to receive 98 507 US$. The payment will help Ever Cat Fuels offset the cost of producing approximately 881 000 gallons of biodiesel at its facility in Minnesota. The company is the first commercial plant to employ the Mcgyan process, and has an annual production facility of 3 million gallons. The USDA noted that the plant, which has been operational for two years, has created 20 full-time jobs. Approximately 90 biodiesel producers nationwide are set to receive payments under the program.

Minnesota was the first state to pass a B2 biodiesel requirement, and this has since increased to B5.[15] The state's required volume of biodiesel is scheduled to rise to B10 by 2012, and B20 by 2015. The NBB is committed to supporting state policy that supports biodiesel. Washington and Pennsylvania both have a B2 requirement in effect. Connecticut, Louisiana, Massachusetts and New Mexico have all passed similar legislations that haven't yet taken effect. The NBB applauded the state of Oregon on its progressive upgrade from a B2 requirement to B5 (the second state to require B5). "Increasing the use of domestically produced, low-carbon fuels like biodiesel is a win-win for Oregon", said Rick Wallace, a senior policy analyst at the Oregon Department of Energy, and the Clean Cities Coordinator of the Columbia–Willamette Clean Cities Coalition. "We're supporting the local economy while reducing pollution, rather than relying entirely on fossil fuels to power our state".[15]

Recently, the NBB rolled out its short-term solution to help restore integrity to the biodiesel renewable identification number (RIN) market, as advised by the RIN Integrity Task Force recently established by NBB chairman Gary Haer and laid out the solution to NBB members in a 'town hall' style meeting last week 14 March 2012.[16] The board did six months of due diligence with Genscape "to develop a program that was accessible and affordable to NBB members, and had the best chance of being accepted by obligated parties and the RIN markets", the NBB said, adding that the system reflects broad-based input from biodiesel producer members and leaders.

9.3.4   Fossil Fuel, GHG Prices and Federal Subsidies

Researchers also predicted the USA biodiesel market penetration using Forest and the Agricultural Sector Optimization Model Greenhouse Gas (FASOMGHG) model.[17] FASOMGHG shows that higher diesel fuel prices translate into higher biodiesel production as the estimated biodiesel production is 5.9 billion gallons in 2030, when the wholesale diesel fuel price is 4 US$ per gallon. FASOMGHG was also used to predict the market penetration of biodiesel given if a market price existed for GHG emissions. The GHG price uses the Intergovernmental Panel on Climate Change (IPCC) 100-year Global Warming Potential (GWP) as an exchange rate between GHGs, and the GWP defines carbon dioxide as equal to 1, methane as 21, and nitrous oxide as 310. The carbon equivalent price is exogenous and ranges from 0 to 100 US$ per metric ton, researchers have shown this price range is effective in reducing GHG emissions.[18] The model predicts the USA aggregate biodiesel production for various carbon dioxide equivalent prices and the wholesale diesel fuel price is set at 2 US\$ per gallon. Higher carbon equivalent prices have a small expansionary impact on the biodiesel industry, because of the competition for carbon credits.

Further, FASOMGHG was used to predict the USA market penetration for biodiesel, if the government did remove the subsidies. Consequently, the federal government subsidies expand the biodiesel industry. If the wholesale diesel price is 1 US\$ per gallon, the industry does not produce any biodiesel. If the diesel price is 4 US\$ per gallon, then FASOMGHG predicts the industry will produce 4 billion equivalent gallons in 2030, resulting in a market penetration of 6%.[17]

In the end it can be concluded that given these trends, and the growing uncertainty surrounding USA oil imports, biodiesel has an important role to play in strengthening USA energy security. Biodiesel will not only help to stretch existing petroleum supplies, but will also help free the USA from the hold of imported oil.

9.4   India

The likelihood of production of biodiesel from edible oil resources in India is almost impossible in view of the country's increasing demand for edible oils. India accounts for 9.3% of world's total oil seed production and contributes as the fourth largest edible oil producing country and even then, about 50% of edible oil is imported for catering the domestic needs.[19,20] Table 9.3 shows the production and consumption of edible oil for India. Figure 9.4 shows the consumption trend of petrol, diesel and total petroleum products in India.[21]

Table 9.3 indicates that India imports about 40–50% of its edible oil domestic requirement and therefore, it is not possible to divert edible oil resources to biodiesel production in this country.

Table 9.3   Consumption of edible oil in India (lakh T).

Year	Availability of oil from domestic sources	Import of edible oil	Consumption of edible oil
2000–01	54.9	41.9	96.8
2001–02	61.5	43.2	104.7
2002–03	47.3	43.7	90.9
2003–04	71.1	52.9	124.04
2004–05	73.1	44.0	117.1
2005–06	78.3	57.9	136.2
2006–07	80.5	64.2	144.7
2007–08	85.4	72.5	157.9

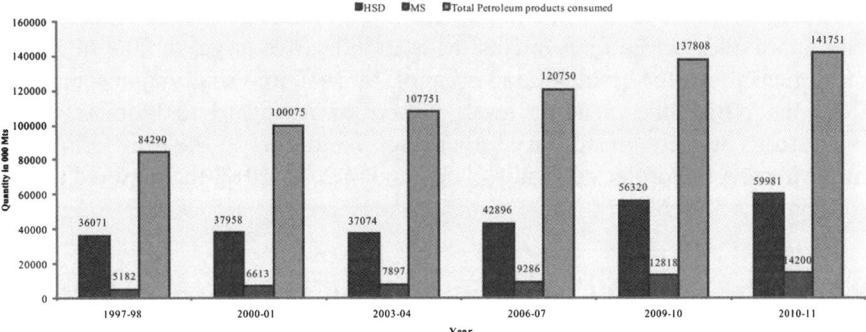

Figure 9.4   Consumption trends of petrol, diesel and total petroleum products in India.

9.4.1   India's Biodiesel Programme

The National Biodiesel Mission (NBM) was started in 2003 for the development of the biodiesel industry.[19] Biodiesel sourced from non-edible seed oils like *Jatropha curcas* and *Pongamia pinnata* seems to be the only possibility in India. A government of India survey showed that out of total land area, 60 Mha are classified as waste and degraded land. India has the third largest road network in Asia with a length of about 3 million km the sides of which can be used for growing jatropha and pongamia crops. India has a railway network of 63 140 km and the land along these tracks can be easily used for cultivation of *J. curcas*. This would also limit soil erosion and improve fertility in addition to oil production.

The government has adopted various policies in order to strengthen the biodiesel sector, such as the selection of appropriate plant varieties available in the wild, developing high-yield varieties and their seedlings, identification of lands and cultivators, cultivation of crops, namely jatropha, collection of seeds and transporting them to processing units, oil extraction, transesterification and production of biodiesel, transporting biodiesel to the depot for blending with petrodiesel and storage, distribution of the blended diesel to retailers and retail selling to the final consumers.[22]

Various government and private organizations such as the Department of Biotechnology, Indian Oil Corporation Ltd, the Aditya Biotech Research Centre (Raipur), the Indira Gandhi Agriculture University (Raipur), the Bhabha Atomic Research Centre (Trombay), the National Botanical Research Institute (NBRI), the Central Salt & Marine Chemicals Research Institute (CSMCRI), leading universities and other R&D institutions have been engaged in biodiesel-related activities like developing appropriate cultivation techniques in terms of standardizing nursery practices (*i.e.*, vegetative/seed/tissue cultures), planting density and planting procedure, fertilization practices, cuttings, seedlings, or other propagation methods, the level of pruning, trimming *etc.*, for different agro-ecosystems.

Anticipating the failure of the NBM, the government of India launched its National Policy on Biofuels in 2008 with an indicative target of 20% blending of biofuels, both for biodiesel and ethanol, by 2017 and as a major departure from the NBM, the blending levels prescribed in regard to biodiesel were intended to be recommendatory rather than mandatory in the near term.[23] In order to meet the projected biodiesel demand in 2017–2018, the required are of wasteland is 18.4 Mha.

9.4.2 Study by the Confederation of Indian Industry and the Ministry of New and Renewable Energy

The study entitled, *Realistic Cost of Biodiesel in India*, conducted by The Confederation of Indian Industry and the Ministry of New and Renewable Energy has determined that the price of jatropha-based biodiesel needs to be increased to create an economically sustainable biodiesel industry within the country.[24] The two organizations ultimately found that the price of jatropha-based biodiesel needs to be increased from 26.5 Indian rupees per L (2.21 US$ per gallon) to 36 rupees per L (3.01 US$ per gallon) under the country's biodiesel policy.

If a B2 blend of jatropha biodiesel is achieved *via* the biodiesel blending initiative during the 2011–2012 timeframe, the study found that India would save approximately 3000 crores (660 US$ million) while generating a revenue of nearly 5500 crores (1.2 US$ billion) in the rural economy, with an annual investment opportunity of 1700 crores (374 US$ million). The study further recommends that the Indian government create a policy framework to make the price of biodiesel self-sustaining. This includes the development of incentives and grants. In addition, the study pointed that a 6000 rupee (132 US$) per ton price for jatropha seeds would ensure that croplands used for food production are not used to produce biodiesel feedstocks while creating a viable business opportunity for the conversion of jatropha seeds into biodiesel. "This will create a pull for biodiesel demand to accelerate the development of the industry, which requires periodic revisits (biannually)", said the organizations in a statement. "These immediate actions will make 'biodiesel blending' a sustainable proposition".

The organizations also stated that the record fluctuations in the price of oil during recent years have created a threat to India's crude oil imports, resulting in a financial burden to the economy, but on the positive side have also created an opportunity to expedite the adaptation and integration of biofuels into the country's energy portfolio. "To counter and mitigate the security threat in dieselized economies like India, biodiesel may be used as a true supplement to fossil diesel with appropriate technological, financial and policy interventions", the organizations continued.

Therefore it can be concluded that successful development of the biodiesel industry in India would also involve appropriate pricing policies for jatropha seeds and biodiesel, and financing policies for organizing cultivation as well as setting up processing factories. Suitable taxation and subsidy policies need also to be formulated in order to provide incentives to the cultivators and other stakeholders.

9.5 Other Parts of the World

9.5.1 Australia

Despite the energy surplus, Australia is a net importer of liquid hydrocarbons and overall energy production continues to exceed its energy consumption, making Australia a significant net energy exporter.[25] In terms of energy sustainability, at current levels of production, Australia's proven reserves of brown coal, black coal and conventional gas are expected to last 500, 100 and 60 years, respectively. Australian reserves of crude oil and condensate represent only a small proportion of total world reserves.

Biofuels capacity for 2011 is an estimated 215 ML of biodiesel and additional plants are currently under consideration and could potentially lift biofuel production to 945 ML by 2014.[25] Supplies of by-products for biodiesel, such as tallow, will likely improve in the future due to improved seasonal conditions and the prospect of fatter slaughter cattle. However, this will likely be balanced by lower slaughter figures. Supplies of waste vegetable oil, the other large feedstock sources for biodiesel, will likely remain largely unchanged.[25]

9.5.2 Brazil

With the successful experience gained in implementing the National Ethanol Production Program (PROALCOOL) during the 1970s, in December 2004 the Brazilian government launched its National Biodiesel Production and Usage Program (PNPB) with the aim of encouraging the introduction of biodiesel into the national energy program.[26] PNPB calls for the introduction of technically, socioeconomically and environmentally sustainable production and use of biodiesel, with a focus on social inclusion and regional development through the generation of employment and income. These guidelines are managed within a regulatory framework that provides tax incentives that favor the inclusion of family farming, the diversification of feedstocks and the development of the poorest regions in the production of biodiesel. The PNPB also quickly generated a set of policies and guidelines, developed and implemented by various ministries (such as those of Finance, Transport, Agriculture, Livestock and Supply, Mines and Energy, Science and Technology, the Environment and Rural Development) to support the strategic guidelines defined in this program.

As of January 2005 (Law11.097/05), refiners and distributors were allowed to add 2% biodiesel to diesel (B2) and in 2008, that percentage became mandatory and due to the rapid response from the supply side, the government

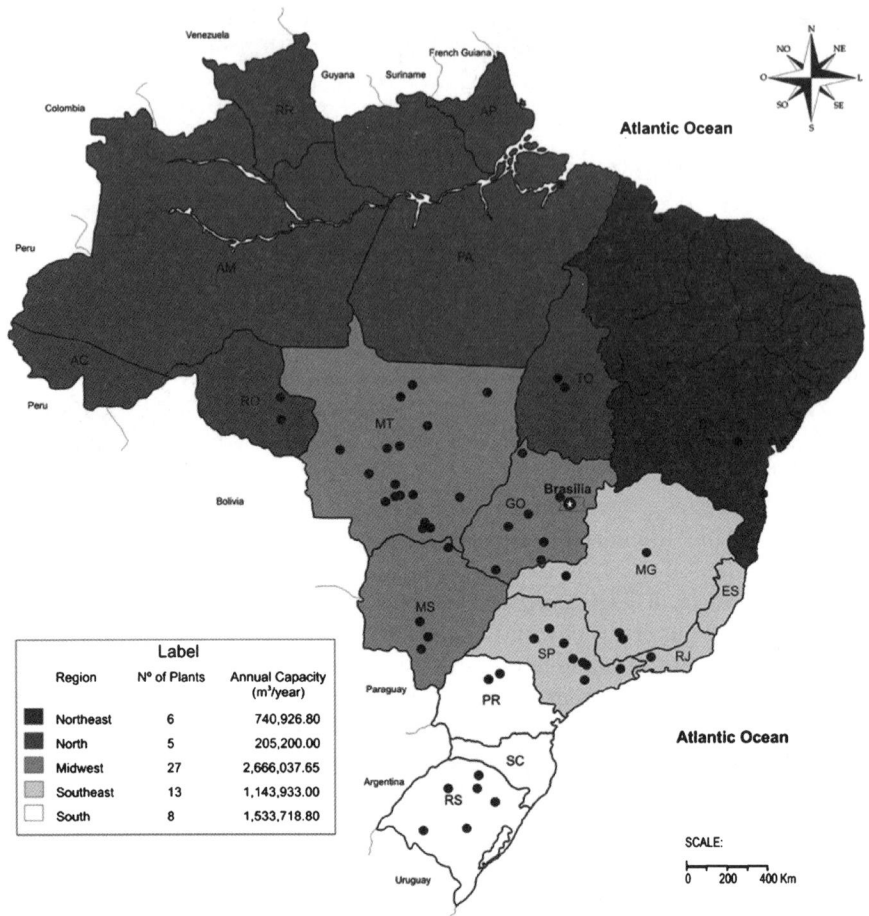

Figure 9.5 Locations of the authorized biodiesel production plants in operation in Brazil.

transformed B5 into a mandatory blending in 2010. Following the implementation of the PNPB, between 2005 and 2011 there was a significant increase in the volume of biodiesel produced in the country, which rose from 736 m^3 in 2005 to 2.39 million m^3 in 2010, making Brazil the world's second largest biodiesel producer.[27] Figure 9.5 shows the location of the authorized biodiesel production plants in operation and Figure 9.6 shows the biodiesel demand forecast in Brazil (1000 m^3 per year for 2006–15.[28,29]

The Social Fuel Seal certificate was started by the ministry of rural development to award biodiesel producers that: (a) acquire minimum percentages of feedstocks from family farmers, (b) enter into contracts with family farmers establishing deadlines and conditions of delivery of feedstocks, and (c) provide technical assistance to the farmers.[27] The Social Fuel Seal offers tax benefits to biodiesel production plants on the condition that part of

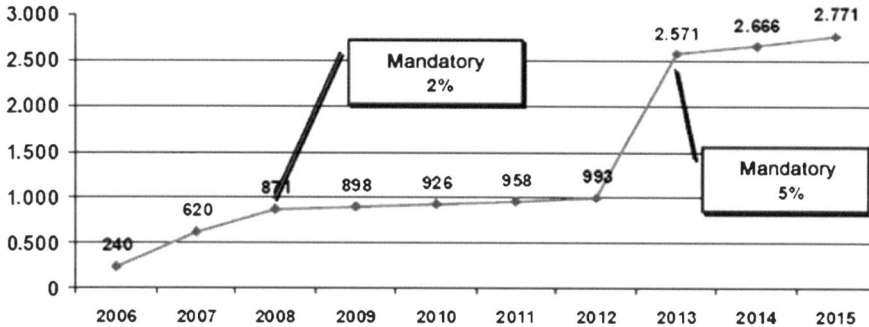

Figure 9.6   Biodiesel demand forecast in Brazil.

the feedstocks (oil seeds) comes from family-based farms thus encouraging their greater participation in the biodiesel production chain. The percentage of the reduction of federal taxes is dependent on the region where the plant is located and the type of feedstock it acquires from the family farms. For example: (a) intensive farming in the north, northeast, and semi-arid regions (castor bean and palm) receives a 30.5% reduction; (b) family-based farming in all regions (any feedstock) receives a 68% reduction; and (c) family-based farming in the north, northeast, and semi-arid regions receives a 100% reduction.

Brazilian biodiesel production is mostly based on soybean, though other important vegetable oil plants are castor bean, palm tree and jatropha.[30] In the short-to-medium term, soy-based biodiesel will correspond to the largest share of Brazil's biodiesel output, in spite of the high cost of this feedstock and the inferior properties of the biofuel compared to mineral diesel. The expansion of large-scale soybean plantations is increasingly menacing the Amazon forest and, therefore, concerns have arisen on the impact of the biodiesel program if soybean is massively used as its feedstock. Therefore, researchers have stressed the possibility of gradually adding alternative feedstocks to the biodiesel supply chain, while enabling the development of catalysts to process more complex oils (acid and enzyme catalysts).[31,32] The Brazilian biodiesel specification is close to the EU and USA specifications (discussed in Chapter 3).

The Brazilian Government, in the process of consolidating the Program and considering the social and environmental benefits of biodiesel will have to design and use policy tools able to define the best ratio of blending, price caps, improvements in the production process, and utilization of the residues.[30] The National Economic and Social Development Bank (BNDES) is providing financial support to investment in biodiesel. One of these measures is a 25% extension in the total loan payoff period for the purchase of machinery that uses at least 20% biodiesel fuel.

Moreover, La Rovere *et al.* have suggested that biodiesel and vegetable oils could be used for decentralized power generation.[30] Besides their use for transport, vegetable oils and biodiesel can also be used to supply electricity to remote communities. Biodiesel production requires either ethanol or methanol as input. In remote communities, with constraints, such as long distances from production centers, small demand, and the lack of roads, ethanol or methanol transportation costs may be significant, therefore in such cases, using vegetable oils for electricity generation would be more useful than its use for biodiesel production. Moreover, the transesterification cost would also be avoided whenever the use of vegetable oil is feasible for power generation.

Researchers further suggested that the Amazon region in Brazil has an enormous diversity of native oil plants, good soil, and climate conditions for high productivity from these crops (*e.g.*, palm oil), besides environmental and social advantages.[30] Pilot units for small-scale generation (below 200 kW) are being tested in some municipalities in the Amazon region, using natural vegetable oil fired in modified engines, such as in Vila Soledade, in the State of Pará, north of Brazil.

There is also scope for a potential synergy between biodiesel and ethanol programs since building a biodiesel transesterification unit integrated to an ethanol distillery may reduce investment costs by 20–25%.[30] Moreover, using ethanol as the input for biodiesel production will also improve technological self-reliance and profitability thanks to increased carbon credits potential.

9.5.3   Canada

The Canadian Environmental Protection Act Bill C-33 mandates a 5–2% renewable content in diesel fuel and heating oil by 2012.[33] According to the Doyletech report, the production of renewable fuels has an aggregate positive impact on the Canadian economy of 2 CAN$ billion annually.[34] The achievement of the 2% biodiesel federal mandate is presumed to require a production capacity of 520 million L of biodiesel by 2012.[35]

Main biodiesel production resources are canola, tallow grease and yellow grease. The relevance of canola in biodiesel production has been increasing and in 2009 it accounted for 14% of the total.[36]

Through the EcoENERGY for Biofuels Program, the maximum incentive rate for biodiesel amounted to 0.20 CAN$ per L and the subsidy will then decrease by CAN$0.04 every year until it is valued at CAN$0.06 in 2016.[37] There are several funding schemes such as the econAGRICULTURE Biofuels Capital Initiative, the Agricultural Bio-Products Innovation Program and Sustainable Development Technology Canada which are in place to expand existing biofuels programs.

9.5.4   China

China's dependency on foreign oil has reached 46.6% and it is one of the world's largest energy consumers, importing 1.63×10^8 tons of crude oil in 2007.[38] Over the next 25 years, 90% of the projected growth in the global energy demand comes from non-core economies and China alone accounts for more than 30%, consolidating its position as the world's largest energy consumer.[39] It is projected that in 2035, China will consume nearly 70% more energy than the USA, which will be the second-largest consumer, even though, by then, per-capita energy consumption in China will still be less than half the level in the USA.

China is the world's second largest GHG emitter.[40,41] Researchers have rightly pointed out that developing bioenergy in China will be an inevitable choice for sustaining economic growth, for the harmonious coexistence of humans and the environment as well as for sustainable development with adjusting energy consumption structure and maintaining national energy security effectively.[42,43]

The experimental study of diesel produced from vegetable oils started in 1981 and presently, several biodiesel production technologies have been developed successfully, using the feedstock of rape oil, *Cormus Wilsoniana* fruit oil, *Jatropha curcas* oil, soybean oil, rice bran oil residue, and restaurant waste oil.[44] Large-scale production of biodiesel started in the early 21st Century and after the instruction by the State Council of the People's Republic of China (PRC) and promotion by the National Development and Reform Commission of the PRC in 2002, large-scale state-owned enterprises, private companies, and international energy enterprises started investing to develop and produce biodiesel.

There are hundreds of biodiesel enterprises in China, and about 60 were over 10 000 ton-class at the end of 2008. Their production capacity exceeds 3.0×10^6 tons per year, but due to inadequate supply of feedstock, many companies are in downtime, and their actual yield was merely 3.0×10^5 tones per year with 4.0×10^5 tons per year of feedstock consumption in 2007.[45]

Researchers have reviewed that from the view of industry life-cycle, the biodiesel industry development in China is transiting into the high-risk, high-profit growth stage from the high-risk, low-profit start-up stage.[44] They further stressed that the one of the important strategies for biodiesel industry development will be to screen, breed and cultivate oil plants sensibly, to explore microalgae resources, to construct a large-scale feedstock supply base, and to explore a sustainable feedstock supply system.

9.5.5 Ghana

In 2006, the Energy Commission, the national agency responsible for energy policy formulation in Ghana, published a detailed Strategic National Energy Plan in four volumes (SNEP I–IV), which charts a road map for energy provision for the country, including biofuels, from 2006–2020.[46] Production of liquid biofuels would mainly affect 17% of the total energy component coming from petroleum products. According to the Bank of Ghana, in 2008 the country imported 2349 US$ million worth of crude oil and refined petroleum products.[47] Ghana has the capacity to increase its palm oil production in order to introduce liquid biodiesel into its energy mix within three years. Increased feedstock production could be achieved initially by simply improving on cultivation methods. The introduction of the new fuels could start with 5% replacement of diesel using palm oil as the feedstock.[46] Sugar cane, jatropha and other non-food alternatives could be developed later as feedstocks. Thermal energy, animal feed and fertilizer could be derived from the biomass resulting from biofuels production.

9.5.6   Indonesia

Indonesia is a developing country and the world's fourth most populous nation, ranking 13th in primary energy use worldwide. The total average increase in population growth between 2000 and 2025 is projected to be 33.2%. Therefore, issues must be addressed by the Indonesian government to overcome any shortage of energy resources in the future.[48] Total annual energy consumption increased from 300 147 GWh in 1980, 625 500 GWh in 1990, 1 123 928 GWh in 2000, to 1 490 892 GWh in 2009, at an annual increase of 2.9%.[49]

In 2007, Indonesia was ranked 15th among fossil-fuel CO_2-emitting nations; emissions from the consumption of oil accounted for 40.9% of its total.[49] Presently, fossil-fuel-based energies such as oil, coal, and natural gas are the major sources of energy in Indonesia. In 2009, oil was the largest single source of energy (48%) followed by natural gas (26%), coal (24%) and renewables (2%). It is expected that total fossil fuel consumption will increase by 52% by 2025. Indonesia's oil, natural gas and coal production is decreasing but consumption is increasing.

Therefore other alternate energy resources such as hydro, wind and biofuels are now being explored. Effort has been put into biodiesel development in Indonesia for more than 10 years.[50] The government of Indonesia announced a National Energy Policy (NEP) in 2006. This policy targets job creation for 3.5 million people, development of 5.25 Mha of biofuel cropland (1.5 Mha palm, 1.5 Mha *J. curcas*, 1.5 Mha cassava, and 750 000 ha sugarcane) on currently uncultivated land, production of 62 000 kL of biodiesel from palm oil (equivalent to 62 000 tons), production of 7.5 million tons of *J. curcas* oil in 2010 and 15 million tons in 2015 on 3 Mha of arable land and development of the biofuel market as an emergency measure against poverty and unemployment from 2006 to 2025.[51] The policy also aims to fulfill an increase in biodiesel blending from 10% of the total consumption of diesel in 2010 to 20% in 2025.[52–55] The biodiesel standard SNI 04-7182-2006 has been approved by the National Standardization Agency (BSN).[56]

Several Indonesian research institutions are working on biodiesel development including Lemigas (Oil and Gas Technology), PPKS Medan (Indonesian Oil Palm Research Institute, Department of Agriculture), ITB (Bandung Institute of Technology), and BPPT (Agency for the Assessment and Application of Technology).[50] In 2006, Pertamina, the state-owned oil and fuel distribution company, as well as the only supplier of biofuels, started selling a blend with 5% biodiesel (B5).

According to Indonesia's Ministry of Energy and Mineral Resources, 520 000 tons of biodiesel were produced in 2007, equivalent to 590 000 kL.[50] In Indonesia, annual biodiesel production increased from 14 500 barrels per day in 2008 to 16 200 barrels per day in 2009 at an average annual increase of 11.7%. It is expected that biodiesel production will increase substantially in Indonesia in the near future due to the availability of bulk biodiesel feedstocks such as palm oil and *Jatropha curcas*.[50] In terms of revenue, production of crude palm oil and crude *Jatropha curcas* oil provides Indonesia with its biggest non-petroleum source of biodiesel, and this is expected to grow in the future.

9.5.7 Iran

Iran is a member of the Organization of the Petroleum Exporting Countries and ranks among the world's top three holders of both proven and natural gas reserves.[57] Iran has an estimated 137.6 billion barrels of proven oil reserves or roughly 10% of the world's total reserves.[58] In 2008, Iran produced 4.2 million barrels of oil per day (bbl/d) equal to about 5% of global production and exported near 2.4 million bbl/d of oil to Asia and EU countries, making it the 4th largest exporter in the world in 2008.

Iran's population has doubled to 70 million people in only 30 years and much electricity is needed for the growing population and economy. Due to infrastructure problems, domestic demands and the economic need to export oil and natural gas, these energy sources cannot fully meet future Iranian electric needs. According to Najafi *et al.*, an energy efficiency program using renewable energy could help cut electric use considerably and meet all the future electricity needs of Iran.[57] The production of renewable energy and energy efficiency products would create thousands of jobs, help the economy, reduces greenhouse gases and save Iranian oil and gas for export.

The same researchers further proved that there is considerable potential for the utilization of algae oil for the production of biofuels in Iran.[57] Geographically Iran has Caspian Sea in the north and Persian Gulf in the south that has natural advantages for algae culturing. Iran has numerous salt lakes that can help scientists advance algae-based biofuel technology. Producing biodiesel from microalgae could ideally replace 5% of the total diesel fuel consumption in a first step. An AB5 (5% algal biodiesel and 95% diesel fuel) could be an optimum fuel for compression–ignition engines since no major engine modifications are required to use biodiesel. If these materials can be efficiently utilized there will be no need to import conventional fuel to the country.

9.5.8 Malaysia

The Malaysian government launched its National Biofuel Policy on 10th August 2005 and the biofuel policy takes into account the development, feasible use, sustainable supply and the spin-off effects of biodiesel in the short, medium and long terms to underscore Malaysia's contribution to the global renewable fuel objective.[59] The implementation of the national policy was spearheaded by the Ministry of Plantation Industries and Commodities of Malaysia. The National Biofuel Policy spurs on the Malaysian biofuel industry. The implementation strategies of the National Biofuel Policy in Malaysia can be categorized into short-, medium- and long-term strategies. On a short-term basis, the Malaysian Standard specifications for B5 will be established. Selected government departments with their fleets of diesel vehicles will participate in trials for using B3 diesel while at the same time, B5 diesel pumps for the public will established at selected fuel stations.

Malaysia was the world's second top producer of palm oil (after Indonesia) in 2008–09. The leading markets for Malaysian biodiesel in 2008 were the USA (71 224 tons), the EU (65 681 tons), Singapore (29 485 tons), South Korea (6500 tons), Romania (3500 tons), Taiwan (3000 tons) and Australia (1200 tons).[60]

Biodiesel producers may be eligible for financial incentives if Pioneer Status (PS) is given to the biofuels sector or by applying for an Incentive Tax Allowance (ITA).[61] The PS scheme allows at least 70% tax reductions on the statutory income obtained from biodiesel production for five years. The ITA plan targets companies with high investment costs in equipment and machinery. The Malaysian government is continuing to revise these policies in order to intensify this huge industry. As a result of intensified production and usage of this renewable energy source, Malaysia could find itself as a main country in this area.

9.5.9 Peru

Peru adopted Law 28054 to promote a biofuels market at the national level in 2003, in line with the government's policy to develop renewable energy resources and as a strategy for poverty alleviation.[62] A regulatory framework on biofuels was established in 2007 setting a blending mandate for ethanol and biodiesel: the mandate stipulates a 2% biodiesel blend with diesel starting in 2009 to be phased up to a 5% biodiesel blend in 2011. A 2% blending mandate will require 1.4 Million of Barrels per day (MBD) of biodiesel and a 5% will require 3.7–4.3 MBD of biodiesel. The government has estimated that the production of oilseed crops in deforested lands can be used to produce the feedstock to meet the blending mandate. The main stress is on palm and jatropha.

Quintero *et al.* have suggested that in Peru, there is a need to promote institutional constructs that support collective action by smallholders so that they can access more of the financial dividends offered by the bioenergy sector.[62] Bioenergy development can create opportunities for the poor through direct and indirect generation of employment that will increase income of families so that these have year-round access to adequate food.

9.5.10 Russia

Russia is naturally rich in fossil fuels and is the second largest oil exporter and the largest natural gas exporter in the world. It also has a large consumer base.[63] *Russia's Energy Strategy* establishes the goal of an efficient, environmentally friendly and financially sound development strategy by 2030.[64] According to *Russia's energy strategy*, the country can save up to 360–400 million tons of oil equivalent (toe), which is about 39–45% of the current annual energy consumption. One of the ways to achieve effective energy management is the development of renewable energy. Russia currently uses only 1% renewable energy from the total amount of its energy consumption.[65] One of the renewable energy options, that is biodiesel, might be obstructed by poor performance in the cold weather. Consumers complain about the reduction of power of the engine and the more frequent maintenance of cars required. Researchers have suggested that despite some technological disadvantages, biofuels can still be used and produced in an efficient way and modern science can find solutions for the improvement of biofuel performance and compatibility.[66] The Minister of Energy of the Russian Federation recently announced that in five years Russia will start using biofuels. Only a few years ago Russia was an outsider in the global biofuel scheme, now the interest in biofuels has grown significantly, therefore policy measures and governmental support should assist further development. Biofuel is a new fuel for the Russian market and it is very important to properly present it to consumers.

9.5.11 Taiwan

Taiwan is a densely populated island with limited natural resources and due to its lack of self sufficient energy resources, Taiwan relies on imports for the majority of its energy.[67] Therefore, exploiting indigenous energy like biodiesel is becoming the need of the hour.

In 1998 at the First National Energy Conference, the Taiwanese government decided to seek new clean energies, and biodiesel development was listed as one of the planned programs. In 2001, the government promulgated the 'Administrative Regulations on the Production and Sales of Renewable Energies Such as Ethanol, Biodiesel, or Oil from Recycled Waste'.[67] To promote the utilization of biodiesel, article 38 of the 'Petroleum Administration Act' explicitly prescribes that producers of renewable energies are exempt from petroleum stockpiling obligations and petroleum fund payments.[68]

In 2001, the Energy and Environment Research Laboratories (EERL) of the Industrial Technology Research Institute (ITRI) associated with the American Soybean Association (ASA) and Taipei City Council, conducted diesel engine performance and emissions measurements on garbage trucks using 100% soy-based biodiesel and a 20% biodiesel blend. In October 2004, EERL cooperated with Taiwan NJC Corporation to establish Taiwan's first biodiesel demonstration system in Chiayi with an annual output of approximately 3000 metric tons. The system uses primarily waste edible oil as its feedstock to produce biodiesel through a transesterification process.[69]

The Environmental Protection Administration (EPA) started the 'Biodiesel Road-Test Program' in garbage trucks in response to the renewable energy development policy of the Executive Yuan, and since 2004 it has provided subsidies on purchasing biodiesel, with total subsidies of 3 US$ million per year.[70] More than 900 garbage trucks have joined in the program using between B10 and B100 biodiesel. This program has helped 13 counties and cities across the nation to collectively consume about 1300 000 L of biodiesel between 2004 and 2005. Road-test results indicate that biodiesel is effective in improving air quality.

In 2005, the Council of Agriculture (COA) selected 90 ha of fallow farmland on which to plant three kinds of energy crops (*i.e.*, sunflowers, rapeseed, and soybeans) as raw materials for producing biodiesel.[71]

Biodiesel development in Taiwan is still in the initial stage. Preliminary assessments show that the annual output of biodiesel generated from waste edible oil could reach approximately 80 000 000 L.[67] To further promote biodiesel in the future, key policy measures such as acquiring material sources, establishing a recycling system, defining economic and legal measures, and improving public acceptance must still be addressed. In addition, the 'Biodiesel Road-Test Program' and the 'Green Energy Development Plan' implemented by the government are expected to continue so as to promote biodiesel gradually. According to the conclusions of the second National Energy Conference, the target of biodiesel promotion is 150 000 000 L by 2020.[67]

9.5.12 Thailand

Due to their over-reliance on imported fuels, the Thai government has instituted a renewable energy policy centering on biofuel use and prioritizing biofuel development as a matter of national interest. Since the start of the biofuel development program over 30 years ago, Thailand has put in place many biofuel policy measures, and the commercialization of biofuels has already begun in the consumer market countrywide.[72] A B2 blend was mandated in all commercial fuel stations in February 2008. Thailand is the only Asian economy to date to embrace biofuels in the main consumer market, where biodiesel blends are available at fuel stations nationwide. As of August 2009, there were 13 licensed commercial biodiesel producers, with a combined production capacity of about 4 460 000 L per day, although the actual production average about 1 500 00 per day. Already over 3400 fuel service stations offer B5.[73]

The range of biofuel measures has included investment promotion, biofuel standards legislation, feedstock productivity development, fuel tax incentives, vehicle specifications and tax incentives, research and development programs, as well as improvements in logistics and transportation networks.

Thailand's current 15-year (2008–2022) alternative energy goals set production targets of biodiesel at 3.0, 3.6 and 4.5 million liters per day in the short-term (by 2011), medium-term (by 2016) and long-term (by 2022), respectively.[74] The biofuel industry in Thailand will remain reliant on subsidies in the near future; however there is significant room for improvement in biofuel cost competitiveness through increased yields.[75] Thailand is expected to import around 200 000 tons of palm oil between 2011 and 2015 in order to meet this envisaged biodiesel share.[76]

9.5.13 Turkey

Turkey is a country which has rich agricultural potential with 23.07 million ha agricultural arable land. 18.11 million ha are currently cultivated and the remaining part is fallow land.[77] Biomass energy ranks second compared to the potential of solar energy; it is 7.9 million tons of oil equivalent (mtoe) in terms of usage and its rate in total energy usage comes first at 13%.[78]

The General Directorate of Electrical Power Resources Survey and Development Administration established the 'Bioenergy Project Group' to follow the topic of biodiesel in Turkey. A pilot-scale biodiesel plant with a capacity of 200 L, which was funded by domestic capital and which has been carrying out experimental production, started production in 2003.[77]

Turkish biodiesel standards are almost same as EN 14213 and EN 14214. The Energy Market Regulatory Authority (EMRA) made it obligatory for biodiesel plants to obtain an operation license and by March 2009, the number of production plants that were granted an operation license by the EMRA was 56. However, the number of active plants has fallen to 3–4 due to the difficulties in legal procedures.[77].

Acaroglu and Aydogan have suggested that Turkey can become an important vegetable oil exporting country by contributing other oilseed crops.[77] After meeting the vegetable oil demands of the country, the rest can be used for biodiesel production. For this reason, sunflower-cultivating areas can be increased in the regions except the Marmara Region. They further suggested that in the Aegean, Mediterranean and South-Eastern Anatolia regions, the farmers could be encouraged to grow a second (sunflower) crop after harvesting of their winter cereals.

9.5.14 Vietnam

Vietnam has issued a Biofuel master plan, which outlines the development of biofuel projects up to 2015, with an outlook to 2025.[72] The government plans to invest 15.3 US$ billion over nine years (2007–2015) and in response to the master plan, many organizations, including governmental offices, educational institutes, private companies, and NGOs, have launched their own initiatives related to their areas of expertise and responsibility. In Vietnam, companies are currently focusing mainly on the feedstock stage of biodiesel production. A pilot plant that produces biodiesel from catfish fat is in operation at a major catfish company, and in addition, a project to produce biodiesel from used cooking oil is being developed.[72]

References

1. http://europa.eu.int/comm/dgs/energy_transport/index_en.html.
2. http://www.ebb-eu.org/stats.php.
3. http://www.ebb-eu.org/pressdl/BlackSeaGrain%20Oct2011.pdf.
4. R. Brand, presented at the Berlin conference on the Human Dimension of Global Environmental Change: Greening of Policies—Policy Integration and Interlinkages, Berlin, 2004.
5. http://www.bmvbw.de.
6. http://www2.biodieselverband.de.
7. http://www.ebb-eu.org/pressdl/Global%20net%20ILUC%2020111013.pdf.
8. http://www.biodieselmagazine.com/articles/8306/obama-talks-energy-policy-during-state-of-the-union-address.
9. www.epa.gov/OUST/fedlaws/publ_109-058.pdf.
10. http://www.biodieselmagazine.com/articles/8310/us-biodiesel-production-surpasses-1-billion-gallons-in-2011.
11. http://www.biodiesel.org/production/production-statistics.
12. http://www.biodiesel.org/news/biodiesel-news/news-display/2012/03/21/biodiesel-production-slips-after-late-2011-growth.
13. http://www.biodieselmagazine.com/articles/8176/dot-awards-grants-for-biodiesel-buses.
14. http://www.biodieselmagazine.com/articles/8078/biodiesel-producers-to-receive-advanced-biofuel-payments.
15. http://www.biodieselmagazine.com/articles/7737/nbb-applauds-oregonundefineds-move-to-b5.
16. http://www.biodieselmagazine.com/blog/article/2012/03/more-details-on-biodiesel-rin-integrity-program.
17. K. R. Szulczyk and B. A. McCarl, *Renewable Sustainable Energy Rev.*, 2010, **14**, 2426.
18. U. A. Schneider and B. A. McCarl, *Environ. Resource Economics*, 2003, **24**, 291.
19. S. Jain and M. P. Sharma, *Renewable Sustainable Energy Rev.*, 2010, **14**, 763.
20. Ministry of Agriculture, *Production of Oil Seeds*, Government of India, New Delhi, 2009.
21. www.ppac.org.in.
22. P. K. Biswas, S. Pohit and R. Kumar, *Energy Policy*, 2010, **38**, 1477.
23. S. Kumar, A. Chaube and S. K. Jain, *Energy Policy*, 2012, **41**, 775.
24. http://www.biodieselmagazine.com/articles/7508/study-recommends-raising-biodiesel-price-in-india.
25. G. Pettrie and M. Darby, Report No. AS1118, USDA foreign agricultural services, Global agricultural information network, Washington, D.C., USA, 2011.

26. http://www.planalto.gov.br/ccivil_03/_ato2004-2006/2005/Lei/L11097.htm.
27. A. D. Padula, M. S. Santos, L. Ferreira and D. Borenstein, *Energy Policy*, 2012, **44**, 395.
28. http://www.anp.gov.br/?pg=29048&m=&t1=&t2=&t3=&t4=&ar=&ps=&cachebust=1323956805646S.
29. http://abag.sites.srv.br/images/pdfs/Gilberto_Ribeiro_de_Carvalho.pdf.
30. E. L. La Rovere, A. S. Pereira and A. F. Simões, *World Dev.*, 2011, **39**, 1026.
31. D. Y. C. Leung, X. Wu and M. K. H. Leung, *Appl. Energy*, 2010, **87**, 1083.
32. D. F. Melvin Jose, R. E. Raj, B. D. Prasad, Z. R. Kennedy and A. M. Ibrahim, *Appl. Energy*, 2011, **88**, 2056.
33. http://www.ecoaction.gc.ca/news-nouvelles/20080626-eng.cfm.
34. www.greenfuels.org.
35. D. Dessureault, Canada Biofuels Annual Report No. CA8057, USDA Foreign Agricultural Service, Washington, D.C., USA, 2008.
36. D. Dessureault, Canada Biofuels Annual Report No. CA9037, USDA Foreign Agricultural Service, Washington, D.C., USA, 2009.
37. G. Sorda, M. Banse and C. Kemfert, *Energy Policy*, 2010, **38**, 6977.
38. http://oilsyggs.mofcom.gov.cn/article/gjhyxw/200801/17511_1.html.
39. http://www.forbes.com/sites/kenrapoza/2011/11/12/what-will-chinas-energy-demand-be-in-two-decades/.
40. National Development and Reform Commission of China, *China's National Climate Change Programme*, National Development and Reform Commission of China, Beijing, China, 2007.
41. L. Zhang, *Geograph. Res.*, 2006, **25**, 1.
42. X. P. Zhou, B. Xiao, R. M. Ochieng and J. Yang, *Renewable Sustainable Energy Rev.*, 2009, **13**, 479.
43. X. H. Wang and Z. M. Feng, *Renewable Sustainable Energy Rev.*, 2004, **8**, 183.
44. Z. Peidong, Y. Yanli, T. Yongsheng, Y. Xutong, Z. Yongkai, Z. Yonghong and W. Lisheng, *Renewable Sustainable Energy Rev.*, 2009, **13**, 2571.
45. http://www.bioon.com/bioindustry/bioenergy/374907_4.html.2008-09-29.
46. G. Afrane, *Energy Policy*, 2012, **40**, 444.
47. Ghana Statistical Service, *Digest of Macroeconomic Data*, Ghana Statistical Service, Accra, Ghana, 2010.
48. http://www.datastatistikindonesia.com/proyeksi/index.php?option=com proyeksi&task=show&Itemid=941&lang=e.
49. http://www.bp.com/productlanding.do?categoryId=6929&contentId=7044622.
50. A. S. Silitonga, A. E. Atabani, T. M. I. Mahlia, H. H. Masjuki, I. A. Badruddin and S. Mekhilef, *Renewable Sustainable Energy Rev.*, 2011, **15**, 3733.
51. www.eaber.org/intranet/documents/96/2114/IEEJ Siang 2009.pdf.
52. H. H. Masjuki, Biofuel Engine: A New Challenge. International & Corporate relations Office (ICR). University of Malaya, Kuala Lumpur, Malaysia. ISBN (978-967-5148-65-1), pp. 1–56, 2010.

53. http://www.agritrade.org/events/documents/Arifin2008.pdf.
54. Indonesia Ministry of Energy and Mineral Resources, *Strategic Plan for New and Renewable Energy of Indonesia*, Indonesia Ministry of Energy and Mineral Resources, Indonesia, 2007.
55. A. Zhou and E. Thomson, *Appl. Energy*, 2009, **86**, 11.
56. http://www.biomass-asia-workshop.jp/biomassws/03workshop/material/papersoni.pdf.
57. G. Najafi, B. Ghobadian and T. F. Yusaf, *Renewable Sustainable Energy Rev.*, 2011, **15**, 3870.
58. http://www.ogj.com/index.html
59. A. Z. Abdullah, B. Salamatinia, H. Mootabadi and S. Bhatia, *Energy Policy*, 2009, **37**, 5440.
60. http://www. bharian.com.my/Current_News/BH/Saturday/BeritaSawit/S
61. R. Hoh, Malaysia Biofuels Annual Report No. MY9026, USDA Foreign Agricultural Service, Washington, D.C., USA, 2009.
62. J. A. Quintero, E. R. Felix, L. E. Rincón, M. Crisspín, J. F. Baca, Y. Khwaja and C. A. Cardona, *Energy Policy*, 2012, **43**, 427.
63. Organisation for Economic Co-operation and Development, *Key World Energy Statistics*, OECD, Paris, 2009.
64. http://www.energystrategy.ru/projects/es-2030.htm.
65. http://minenergo.gov.ru/activity/vie/.
66. N. Lykova and J. E. Gustafsson, *Scientific J. Riga Technical University*, 2010, **4**, 64.
67. Y. Huang and J. Wu, *Renewable Sustainable Energy Rev.*, 2008, **12**, 1176.
68. The Legislative Yuan, *Petroleum Administration Law*, Beijing, 2001.
69. C. W. Lan, H. P. Wan, H. C. Chen, H. T. Lee and W. J. Lu, present at the 28th Annual IAEE International Conference, Taipei, 2005.
70. Environmental Protection Administration, *Report on Promoting Biodiesel in Taiwan*, EPA, Beijing, 2006.
71. J. H. Wu, presented at the Taiwanese Symposium on Biofuel and Energy Crops, Taipei, 2005.
72. S. Sukkasi, N. Chollacoop, W. Ellis, S. Grimley and S. Jai, *Renewable Sustainable Energy Rev.*, 2010, **14**, 3100.
73. http://www.dede.go.th/dede/index.php?id=351.
74. http://www.dede.go.th/dede/fileadmin/upload/nov50/ mar52/REDP_present.pdf.
75. National Center for Genetic Engineering and Biotechnology, *Feasibility Study of Increase Production of Sugarcane, Cassava and Oil Palm for Biofuels Production*, NCGEB, Bangkok, 2009.
76. S. Preecharjarn, GAIN Report No. TH0079, USDA Foreign Agricultural Service, Washington, D.C., USA, 2010.
77. M. Acaroglu and H. Aydogan, *Biomass Bioenergy*, 2012, **36**, 69.
78. World Energy Council Turkish National Committee, *Turkish Energy Report 2007–2008*, WECTNC, Ankara, Turkey.

CHAPTER 10

The Food Versus Fuel Issue: Possible Solutions

10.1 Introduction

It is currently under international discussion, if and to what extent, biofuel production affects feedstock prices. Are biofuels solely responsible for enhancing food prices? Are these concerns really genuine and if yes, then to what extent? This chapter will provide the answer to this question and also provide possible solutions related to the so-called 'food *versus* fuel issue'.

10.2 The Issue

According to an analysis by the International Food Policy Research Institute (IFPRI), bioenergy accounted for 30% of the escalation in global cereal prices between 2000 and 2007, and for nearly 40% of the increase in the real global price of maize during that time.[1] The growing use of food crops to produce biofuels was considered by the former Special Rapporteur on the Right to Food of the UN Human Rights Council as: "A crime against humanity".[2] The IFPRI also modeled the possible effects of continued development of biofuels and found that stress on regional water supplies would increase only marginally.[3] Researchers investigated the potential habitat and biodiversity losses that may result from an increase in global biodiesel production capacity to meet future biodiesel demands (an estimated 277 million tons per year by 2050) and estimated substantial increases in cultivated areas for all major biodiesel feedstocks, including soybean in the USA (33.3–45.3 million ha), sunflower seed in Russia (25.7–28.1 million ha), rapeseed in China (10.6–14.3

Biodiesel: Production and Properties
Amit Sarin
© Amit Sarin 2012
Published by the Royal Society of Chemistry, www.rsc.org

million ha), and oil palm in Malaysia (0.1–1.8 million ha) (Figure 10.1).[4] It was stressed that, because soybean and oil palm are most intensively cultivated in biodiversity hotspots, any future intensification of soybean or oil palm production, without proper mitigation guidelines, will likely further threaten the high concentrations of globally endemic species in these areas. Minimum estimates were made assuming that 50% of existing arable and permanent cropland in the country will be converted to biodiesel feedstock before non-agricultural lands are converted. Maximum estimates were made assuming all expansion of feedstock agriculture will occur in non-agricultural lands.

Several non-governmental organizations have accused oil palm growers in Southeast Asia of destroying large tracts of tropical forests and threatening the survival of many native species, including the orangutan (*Pongo pygmaeus*).[5] But producers argue that oil palm cultivation is not a threat to biodiversity because only disturbed forests or pre-existing croplands have been converted to oil palm with minimal disturbance to pristine habitats.

Ajanovic found that in spite of the increasing use of wheat and corn for biofuel production, their prices were relatively stable in the period 1996 to 2006. In the period between 2006 and 2008 these commodity prices increased by more than 50%.[6] But after the price spike in 2008, in July 2009 corn and wheat prices had declined to the same level as in 2006, as well as 1996. This proves that the volatility of commodity prices is linked to the development of the crude oil price. The oil price has risen continuously since about 2002, but with the drop of in oil price after 2008, feedstock prices also decreased significantly. Hence, there is a clear correlation between oil prices and the

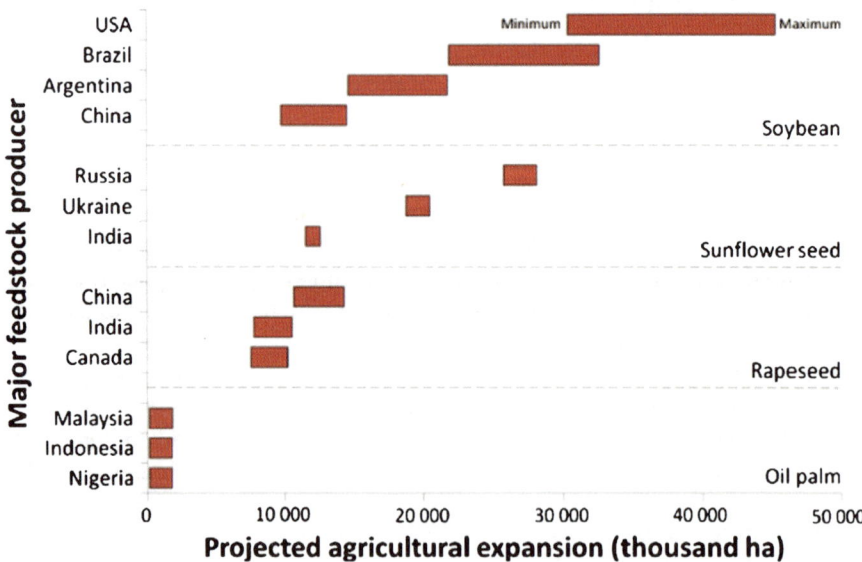

Figure 10.1   Projected agricultural expansion in major biodiesel feedstock producers to meet global biodiesel demand by 2050.

prices of agricultural commodities. Ajanovic further concluded that oil price increases impact feedstock prices not only because of transportation expenses, but also in terms of how much farmers have to spend on oil and the price of fertilizer because energy costs are a significant part of fertilizer, farming and food distribution costs. Therefore the argument, that increases in biofuels production have a significant impact on feedstocks prices, does not hold.

Another available analysis suggests that biofuel production had a modest (3–30%) contribution to the increase in commodity food prices observed up to mid-2008.[7]

10.3 Possible Solutions

As discussed in the previous section, researchers have proved that the production of biofuel has had only a modest influence on food prices. The underlying causes of rising food prices are many and complex, such as adverse weather conditions that affect crop productivity, speculative or precautionary demand for food commodities, and inappropriate policy responses such as export bans on foods.[8,9] Any influence of biofuel on food prices may not be significant. Therefore in this section we will discuss the various solutions to the food *versus* fuel issue. In this section, problems related to both biodiesel and ethanol are discussed, because of the obvious reason that land is used for production of both substances.

10.3.1 Use of Optimum Blends of Biodiesels

Sarin *et al.* suggested that as production of palm biodiesel is in excess in some countries, whilst in other countries research is focused on jatropha biodiesel, jatropha biodiesel, could be blended with palm methyl ester.[10] Jatropha biodiesel has poor oxidation stability with good low-temperature properties and on the other hand, palm biodiesel has good oxidative stability, but poor low-temperature properties. The combination of jatropha and palm produces an additive effect on these two critical properties of biodiesel. Since palm biodiesel has poor low-temperature properties like cloud point and pour point, the blending of jatropha biodiesel would improve these. Therefore, optimum mixture of jatropha biodiesel with palm biodiesel can lead to a synergistic combination with improved oxidation stability and low-temperature properties. This techno-economic combination of jatropha and palm biodiesel could be an optimum mix for Asian energy security.[10] Researchers found the same result when blending palm biodiesel synthesized from edible oil with jatropha and pongamia biodiesel respectively, and both jatropha and pongamia biodiesels remarkably improved the low-temperature flow properties of palm biodiesel, so its use can be minimized.[11,12] Palm biodiesel also improved the oxidation stability of jatropha and pongamia biodiesel.[13]

This solution, through the international treaties, can be used to solve the food *versus* fuel issue. It can be concluded that the feedstock for the synthesis of biodiesel must have a suitable combination of saturated as well as unsaturated fatty compounds to achieve improved oxidation stability and low-temperature properties.

10.3.2 Development of Resources of Biofuel that Require Less Arable Land

Research should also be focused towards the use of feedstocks like macroalgae, perennial grasses, wood, or municipal waste for biofuel production. Most of these crops can be cultivated on marginal or agriculturally degraded lands, and thus may not compete with food production. High-diversity mixtures of grassland species can even provide greater bioenergy yields and greenhouse gas reductions than certain conventional bioethanol or biodiesel production systems.[14]

Campbell found that macroalgae is another potential source of biofuel feedstock. Aquatic unicellular green algae, such as *Chlorella spp.*, are typically considered for biodiesel production owing to their high growth rate, population density, and oil content.[15] Algae have a much higher productivity (90 000 L of biodiesel per hectare) than soybean (450 L per hectare), rapeseed (1200 L per hectare), or oil palm (6000 L per hectare) and macroalgae cultures are not land-intensive.[15,16]

Waste biomass like wheat straw, wood pieces leftover after timber extraction, and municipal wastes (*e.g.*, waste paper, waste food scraps, used cooking oils) can also be used for the production of biofuel. A study estimated that a city of one million people could provide enough organic waste (1300 tons per day) to produce 430 000 L of bioethanol a day, which could meet the needs of about 58 000 Americans, 360 000 French, or 2.6 million Chinese at current rates of per capita fuel use.[17] Koh *et al.* estimated that the 50 000–156 000 tons of horticultural biomass collected each year from about 1 million planted trees in Singapore could be used to produce 14–58 million L of bioethanol that could displace 1.6–6.5% of the country's transport gasoline demand.[18]

The large-scale use of biofuels will probably not be possible unless second-generation technologies based on ligno-cellulosic biomass that require less arable land can be developed commercially.[19] Tirado *et al.* suggested that further investment is needed in developing technologies to convert cellulose to energy, which could provide smallholders with a market for crop residues.[20]

10.3.3 Use of Next-Generation Technologies

There is also a need to develop process technologies that convert these next generation feedstocks to liquid fuels. Biotechnology may also determine the future role of biofuels.[21] Advances in plant genomics could lead to the production of higher yielding biofuel crops, reducing both land requirements and energy input, which may reduce land-use conflicts and greenhouse gas emissions.[22] Vinocur and Altman suggested that biofuel crops may also be genetically engineered to be more resistant to pests, diseases, or abiotic stresses (*e.g.*, drought), which would provide a stable supply of feedstock.[23] Therefore these methods can reduce the land area requirement for biofuels and alleviate pressure on both natural habitats and land for food production.

10.3.4   Appropriate use of Wasteland

Appropriate policies can make biofuel development more pro-poor and environmentally sustainable.[20] For example, poor farmers might be able to grow energy crops on degraded or marginal land not suitable for food production.

The former chairman of Archer Daniels Midland, G. Allen Andreas, said "I think any knowledgeable person in today's world would recognize the fact that the reason we've got malnutrition and hunger is not because we're turning food into fuel. We've got hundreds of millions of acres of land in Brazil that are suitable for arable development into farmland that still have not been cultivated without any infringement on the environment. There's plenty of capacity to make food".[24]

Another study concludes that there is potential for the development of biofuels in Botswana without adverse effects on food security, due mainly to availability of idle land which accounted for 72% of agricultural land in the eastern part of the country in 2008.[25] It is suggested that farmers could be incentivized to produce energy crops and more food from such land. The use of marginal and idle land for biofuel production is therefore suggested as a solution for reducing these adverse impacts.

There are large tracts of wastelands in India, which have been lying almost barren for decades and it has been observed that around 13.4 Mha of wasteland as identified by the Indian Government could be utilized for biofuel plantations using wasteland afforestation.[26]

10.3.5   Development of Codes of Conduct

Codes of conduct could be developed that will constrain excessive exuberance on cultivating biofuels, as with fair trade guidelines, or ensure that cultivation occurs in a sustainable manner.[20] For example the Roundtable on Sustainable Biofuels has developed a third-party certification system for biofuel sustainability standards, encompassing environmental, social and economic principles and criteria for which biofuel producers can apply.[27] For only those fuels that have environmental certification would be allowed to be commercialized in the market. With the implementation of these measures, an appropriate regulatory framework will be established for the future growth and expansion of crops that can be synthesized as biofuels, hence reducing the environmental risks related to their production and assuring the necessary land quotas for food production.

10.3.6 Development of Pro-Poor and Environmentally Friendly Policies

In order to assure that biofuel development is pro-poor, environmentally friendly and supports food security and nutrition, Tirado *et al.* suggested a number of steps such as: developed-country governments should remove trade barriers to developing-country biofuel exports; developed-country governments, international organizations and international financial institutions should provide financial and technical assistance to pro-poor sustainable biofuel projects in developing countries; developing-country governments need to conduct food and nutrition security impact assessments before launching biofuel development projects; developing-country government policies should make opportunities available to smallholders, including women farmers, to participate in biofuel production, such as incentives to encourage outgrower schemes and labor-intensive processing plants; policies should also encourage technology spillovers from biofuel production that can enhance food crop production and policies should favor production of biofuel crops with a small environmental footprint that can contribute to climate change adaptation and mitigation strategies.[20]

Ajanovic has suggested that the farmers need a certain market price level to have an incentive to grow feedstock. Hence, a more intensive competition due to feedstock use for biofuels could finally lead to an over-all healthier market.[6]

Policies directed toward setting efficient market prices should be implemented. These policies will allow the market to adjust to increasing biofuel demands and include allowing free markets to adjust to changes in crop usage, constant infusion of public-sponsored agricultural research and outreach, a shift to sustainable perennial crops arresting topsoil erosion, on the demand-side improving energy efficiency, for each particular country's food and fuel imports.[28] There may be short-run policies that include expanding emergency humanitarian assistance to food-insecure areas, complete negotiations on reducing agricultural trade restrictions, create precautionary agricultural commodity buffer stocks, establish a global food-price monitoring system, and educating consumers to expect greater food-price volatility.[29]

Another opportunity for wider welfare benefits may be through the use of small-scale biofuel production models that convert feedstocks locally.[30] Small-scale biofuel production will generate employment and income opportunities in the growing and processing of feedstocks; however, since schemes are introduced to satisfy a range of community needs, the benefits of the technology can be felt locally. However, strong government support for small producers will be necessary in order to ensure that biofuel production in developing nations is sustainable and brings welfare benefits to rural areas. The public sector needs to set the legal, fiscal, and institutional framework for biofuel production and there should be public–private partnerships because these can help to ensure that supply chains generate income and employment for small producers and laborers.[31] The private sector can play a critical role in technology transfer and related capacity building, especially if they create the kind of technology spillovers that could improve the productivity of smallholder agriculture.[32]

There are many incentives that can be offered by governments to increase the development of the biodiesel industry and maintain its sustainability, such as crop plantation in abandoned and fallowed agricultural lands, implementation of carbon taxes, subsidizing the cultivation of non-food crops, and exemption from oil taxes.[30,33]

Therefore, comprehensive assessments of energy policies will be required to avoid potential conflicts with food production, land change, and provide the most benefits. Governments should also enhance the promotion of the biodiesel industry to increase public awareness of the biodiesel industry, which will help governmental biodiesel policies to garner sufficient public support.[34] The international treaties to make the optimum blend of biodiesel can be one of the possible solutions of this food *versus* fuel issue.

References

1. http://www.ifpri.org/blog/biofuels-and-grain-prices-impacts-and-policy-responses.
2. G. Ferrett, *BBC News*, 27 October 2007.
3. M. W. Rosegrant, T. Zhu, S. Msangi and T. Sulser, *Review Agri. Econ.*, 2008, **30**, 495.
4. L. P. Koh, *Conserv. Biol.*, 2007, **21**, 1373.
5. L. P. Koh and D. S. Wilcove, *Nature*, 2007, **448**, 993.
6. A. Ajanovic, *Energy*, 2011, **36**, 2070.
7. S. A. Mueller, J. E. Anderson and T. J. Wallington, *Biomass Bioenergy*, 2011, **35**, 1623.
8. http://www.adb.org/Documents/reports/food-prices-inflation/foodprices-inflation.pdf.
9. ftp://ftp.fao.org/docrep/fao/010/ai465e/ai465e00.pdf.
10. R. Sarin, M. Sharma, S. Sinharay and R. K. Malhotra, *Fuel*, 2007, **86**, 1365.
11. A. Sarin, R. Arora, N. P. Singh, R. Sarin, R. K. Malhotra and K. Kundu, *Energy*, 2009, **34**, 2016.
12. A. Sarin, R. Arora, N. P. Singh, R. Sarin, R. K. Malhotra and S. Sarin, *Energy Fuels*, 2010, **24**, 1996.
13. A. Sarin, R. Arora, N. P. Singh, R. Sarin and R. K. Malhotra, *Energy*, 2010, **35**, 3449.
14. D. Tilman, J. Hill and C. Lehman, *Science*, 2006, **314**, 1598.
15. M. N. Campbell, *Guelph Eng. J.*, 2008, **1**, 2.
16. A. L. Haag, *Nature*, 2007, **447**, 520.
17. http://cgse.epfl.ch/webdav/site/cgse/shared/Biofuels/BiofuelsforTransport-WorldWatchInstituteandtheGermanGovernment.pdf.
18. L. P. Koh, H. T. W. Tan and N. S. Sodhi, *Science*, 2008, **320**, 1419.
19. http://www.iea.org/textbase/nppdf/free/2006/weo2006.pdf

20. M. C. Tirado, M. J. Cohen, N. Aberman, J. Meerman and B. Thompson, *Food Res. Int.*, 2010, **43**, 1729.
21. E. Kintisch, *Science*, 2008, **320**, 478.
22. http://www.iea.org/textbase/nppdf/free/2004/biofuels2004.pdf.
23. B. Vinocur and A. Altman, *Curr. Opin. Biotechnol.*, 2005, **16**, 123.
24. http://today.reuters.com/news/articlebusiness.aspx?typeconsumerProducts& storyIDnN02286601&frombusiness, 2007
25. D. L. Kgathi, K. B. Mfundisi, G. Mmopelwa and K. Mosepele, *Energy Policy*, 2012, **43**, 70.
26. S. Kumar, A. Chaube and S. K. Jain, *Renewable Sustainable Energy Rev.*, 2012, **16**, 1089.
27. http://cgse.epfl.ch/webdav/site/cgse/shared/Biofuels/Version%20One/ Version%201.0/09-11-12%20RSB%20PCs%20Version%201.pdf.
28. M. Wetzstein and H. Wetzstein, *Energy Policy*, 2011, **39**, 4308.
29. C. Qiu, G. Colson and M. Wetzstein, in *Biofuel*, ed. M. Bernardes, InTech, Rijeka, Croatia, 2011.
30. M. Ewing and S. Msangi, *Environ. Sci. Policy*, 2009, **12**, 520.
31. J. Woods, *Bioenergy and Agriculture: Promises and Challenges*, International Food Policy Research Institute, Washington DC, 2006.
32. C. Arndt, R. Benfica, F. Tarp, J. Thurlow and R. Uaiene, Biofuels, Poverty, and Growth: A Computable General Equilibrium Analysis of Mozambique, International Food Policy Research Institute, Washington DC, 2008.
33. J. C. S. Wassell and T. P. Dittmer, *Energy Policy*, 2006, **34**, 3993.
34. S. Lim and L. K. Teong, *Renewable Sustainable Energy Rev.*, 2010, **14**, 938.

Subject Index

Note: A page reference in italics indicates that a topic appears on that page only in a Figure or Table.